SpringerBriefs in Mathematical Physics

Volume 51

W0235154

SpringerBriefs are characterized in general by their size (50–125 pages) and fast production time (2–3 months compared to 6 months for a monograph).

Briefs are available in print but are intended as a primarily electronic publication to be included in Springer's e-book package.

Typical works might include:

An extended survey of a field
A link between new research papers published in journal articles
A presentation of core concepts that doctoral students must understand in order to make independent contributions
Lecture notes making a specialist topic accessible for non-specialist readers.

SpringerBriefs in Mathematical Physics showcase, in a compact format, topics of current relevance in the field of mathematical physics. Published titles will encompass all areas of theoretical and mathematical physics. This series is intended for mathematicians, physicists, and other scientists, as well as doctoral students in related areas.

Springer Briefs in a nutshell
SpringerBriefs specifications vary depending on the title. In general, each Brief will have:
50 – 125 published pages, including all tables, figures, and references
Softcover binding
Copyright to remain in author's name
Versions in print, eBook, and MyCopy

Katsushi Ito · Hongfei Shu

ODE/IM Correspondence and Quantum Periods

 Springer

Katsushi Ito
Department of Physics
Institute of Science Tokyo
Tokyo, Japan

Hongfei Shu
School of Physics
Institute for Astrophysics
Zhengzhou University
Zhengzhou, Henan, China

ISSN 2197-1757   ISSN 2197-1765 (electronic)
SpringerBriefs in Mathematical Physics
ISBN 978-981-96-0498-2   ISBN 978-981-96-0499-9 (eBook)
https://doi.org/10.1007/978-981-96-0499-9

This Springer imprint is published by the registered company Springer Nature Singapore Pte Ltd.
The registered company address is: 152 Beach Road, #21-01/04 Gateway East, Singapore 189721, Singapore

If disposing of this product, please recycle the paper.

Preface

Integrability plays an important role in solving a problem in physics and mathematics. Integrability means that there exist enough conserved charges in the system. For a classical dynamical system with a finite number of degrees of freedom, these conserved charges determine the solutions of the equations of motion completely. Integrability also helps us to solve the quantum system with infinite degrees of freedom. An infinite number of conserved charges in a quantum field theory restricts the scattering matrices of the particles. In particular, the $(1+1)$-dimensional quantum system is exactly solvable when the scattering matrix obeys certain consistency conditions: the Yang–Baxter equations. From the Yang–Baxter equations, one can study the model through the Bethe ansatz equations, the commuting transfer matrices, and the factorization of S-matrices, which determine the energy spectrum and its thermodynamic properties.

A physical system in nature does not always have such integrability, and we need some approximation methods to understand its properties. Let us consider the case of the Schrödinger equation of a one-dimensional system. The Schrödinger equation has an irregular singularity at infinity, which means the power series solution around infinity is an asymptotic series. The solutions change discontinuously according to the change of direction in the complex domain, which is called the Stokes phenomenon. The Stokes coefficients that connect the asymptotic solutions and the monodromy coefficients that relate the solutions around the regular singularities are strongly constrained by the consistency conditions for the solutions. The inverse problem of finding the solution from the Stokes and monodromy data is called the Riemann–Hilbert problem. Ecallé applied the Borel resummation method to the asymptotic series and formulated the Stokes phenomena by the Alien calculus. The method is called resurgence, which is a powerful approach to studying the asymptotic series. Voros applied the resurgence method to the WKB (Wentzel–Kramers–Brillouin) analysis of the Schrödinger equation with double-well potential and solved the energy spectral problem.

The above two subjects look very different at first sight. However, Patrick Dorey and Roberto Tateo in 1998 showed that there is a relation between them, where the spectral function for the Schrödinger equation and the Y-functions in the quantum integrable models share the same functional relations with the same analyticity and asymptotics. This surprising link is called the ODE/IM (ordinary differential equation/integrable model) correspondence. This is a different type of relation between the classical and quantum integrable systems. It turns out to have many applications in physics and mathematics.

These integrability structures are not only for two-dimensional systems but also appear in different theories in higher dimensions. Thus the integrability-based approach is powerful to solve problems in higher dimensions. One of the reasons to share the same or similar integrability structure is the duality in quantum field theories/superstring theories. Duality implies that a single physical system can be described by some different setups, where two descriptions can be complementary to each other, in the sense that they cover the different regions of the parameter space of the original system. For example, the AdS/CFT (anti-de Sitter/conformal field theory) correspondence predicts the equivalence between gauge and superstring theories in a curved background. One can calculate the physical observables in gauge theories from the string theories which are regarded as the classical integrable models. Moreover, the low-energy effective theory of supersymmetric gauge theories is characterized by the periods of the Riemann surface, which is the spectral curve associated with the classical integrable system. The relation to the gauge theories provides a new understanding of the ODE/IM correspondence and quantum integrability. The key objects for these theories are variously referred to as the WKBperiods, the quantum SWperiods, the energy of the pseudo-particles, the Voros symbols, the Fock–Goncharov coordinates, etc. Here we simply call them the quantum periods.

The purpose of this monograph is to review the basic concepts of the ODE/IM correspondence and its application to quantum mechanics. We want to explain the relations among these theories by using simple examples. We focus on the analysis from the ODE side mainly since the explanation of the integrable models takes many pages and grows beyond the limit of this monograph.

Tokyo, Japan   Katsushi Ito
Zhengzhou, China   Hongfei Shu

Acknowledgments

We would like to thank Zoltan Bajnok, Janos Balog, Davide Fioravanti, Daniele Gregori, Yasuyuki Hatsuda, Saki Koizumi, Takayasu Kondo, Kohei Kuroda, Si Li, Yong Li, Christopher Locke, Yongchao Lü, Marcos Mariño, Takafumi Okubo, Hao Ouyang, Marco Rossi, Kazuhiro Sakai, Yuji Satoh, Junji Suzuki, Roberto Tateo, Gabor Zsolt Toth, Dan Xie, Jingjing Yang, Mingshuo Zhu, Rui-Dong Zhu, Peng Zhao, and Hao Zou for useful discussions. The work of K.I. is supported in part by Grant-in-Aid for Scientific Research 21K03570, 18K03643 from Japan Society for the Promotion of Science (JSPS). The work of H.S. is supported in part by the National Natural Science Foundation of China No. 12405087, the Startup Funding of Zhengzhou University (Grant No. 121-35220049), JSPS Research Fellowship 17J07135 for Young Scientists from Japan Society for the Promotion of Science (JSPS), the grant "Exact Results in Gauge and String Theories" from the Knut and Alice Wallenberg Foundation and the Beijing Postdoctoral Research Foundation. Last but not least, H. S. would like to thank his parents (Guoying Shu and Qiaozhi Wang) and his wife (Shixiao Sun) for their unwavering support and constant encouragement.

Tokyo, Japan   Katsushi Ito
Zhengzhou, China   Hongfei Shu

Acknowledgments

Contents

Chapter 1
ODE/IM Correspondence

The spectral problem of the Schrödinger equation is one of the most fundamental problems in quantum mechanics. Since the exact solution can be found in a few examples, one can use perturbative methods to obtain the energy spectrum when the expansion parameter is small, which is expected to approach an exact value. However, the perturbative expansion provides an asymptotic series sometimes and hence is divergent for any value of the expansion parameter. Therefore, the treatment of the series requires a specific mathematical procedure.

The WKB (Wentzel–Kramers–Brillouin) analysis [1–3], which is an expansion of the wave function in the Planck constant $\hbar$, provides an example of asymptotic series. Based on the Borel resummation technique, André Voros studied the WKB approach to the quadratic potential exactly in $\hbar$ [4]. He studied the analytical structure of the spectral functions of the Schrödinger operator, which contains the perturbative and non-perturbative information of the spectrum. In particular, the spectral functions are related to the Stokes multipliers of the asymptotic problem of the Schrödinger equation and are shown to satisfy certain functional relations. The mathematical details of the asymptotic analysis of the second-order differential equation have been developed by Yasutaka Sibuya [5], where the functional relations satisfied by the Wronskians of the solutions were obtained.

In 1998, Patrick Dorey and Roberto Tateo observed that the spectral determinant of the quartic oscillator in quantum mechanics satisfies the same functional relations as the A_3-related Y-system appearing in the integrable perturbation of two-dimensional conformal field theory [6]. They also observed similar coincidences for further examples of higher-order potentials and conjectured that the x^{2M} anharmonic oscillators are related to the A_{2M-1} Y-systems. This "surprising link" [6] is named the ODE/IM(ordinary differential equation/integrable model) correspondence in the review paper [7]. From the viewpoint of the integrable model, the spectral determinants are nothing but the Q-functions introduced by Baxter to solve the spectrum of the Hamiltonian of the eight-vertex model [8]. In [9, 10], the T-functions, which

© The Author(s), under exclusive licence to Springer Nature Singapore Pte Ltd. 2025   1
K. Ito and H. Shu, *ODE/IM Correspondence and Quantum Periods*, SpringerBriefs in Mathematical Physics 51, https://doi.org/10.1007/978-981-96-0499-9_1

give the eigenvalues of the transfer matrices of the integrable models, are shown
to correspond to the Wronskians of the asymptotically decaying solutions for the
Schrödinger equations with the centrifugal term in addition to anharmonic potential.
Baxter's T-Q relations can be regarded as the connection formula of the solutions
of the Schrödinger equation at the origin and the infinity. These relations are found
to take the same form as those of the six-vertex model or spin 1/2 XXZ model. A
standard procedure in the integrable model tells us how to construct the Y-functions
from the T-functions, from which one can derive the thermodynamic Bethe ansatz
(TBA) equation. One can calculate the free energy and the effective central charge
of the system using the TBA equation. The results from the ODE and the IM show
perfect agreement. This also implies the original spectral problem can be solved by
the integrable model.

The method based on the functional relations in the quantum integrable models is powerful analytically and also numerically. To compute the spectrum of the
Hamiltonian of the two-dimensional classical lattice model or the one-dimensional
quantum spin chain model, integrability greatly helps, where conserved charges
restrict the model strongly. Sometimes integrability requires self-consistent relations
among the quantities, which gives the exact information of the model. For example,
the Bethe ansatz equations for the spectrum are highly non-linear algebraic equations. These equations can be solved by rewriting the equations into the integral
equation called the NLIE [11, 12]. Furthermore, the Y-system [13] and the T-system
[14] satisfied by the T-/Y-functions have profound mathematical structure and many
applications [15]. The ODE/IM correspondence derives these functional relations
from a different viewpoint.

In this chapter, we will explain the ODE/IM correspondence for the Schrödinger
type ordinary differential equation with monomial potential following [6]. We obtain
some functional relations which characterize the consistency between the asymptotic
solutions of the ODE. We will check this correspondence between the ODE and the
IM numerically. The functional relations also appear in the context of the integrable
models, which can be seen in [7]. It is also reviewed in the Appendix of this book.

1.1   ODE/IM Correspondence for the Second-Order ODE

Let us consider the second-order ODE

$$\left(-\frac{d^2}{dx^2} + x^{2M} - E \right) \psi(x) = 0, \tag{1.1}$$

which is defined in the complex x-plane. This is the stationary Schrödinger equation
of a particle with mass $m = 1/2$ in a homogeneous potential x^{2M} and energy E with
$\hbar = 1$. Here we regard E as a complex parameter. The parameter M is assumed to
be real and positive. We will discuss the case where $2M$ is a positive integer. There
are a few examples such that Eq. (1.1) is solved exactly. For $M = 1/2$, the solution

is represented by the Airy function. For $M = 1$, the ODE becomes the Schrödinger equation with harmonic potential, where the solution with appropriate boundary condition is expressed in terms of the Hermite polynomials. In these cases, we can study the energy spectrum based on the solutions. However, for general M, we do not know the exact solution of this ODE represented by special functions. We will apply the WKB method to study the spectral problem for Eq. (1.1).

The ODE (1.1) is regular in the complex plane, where we can investigate the power series solutions around some points in the complex plane. However, the power series around the infinity provides an asymptotic series, which has a zero convergence radius. Such a point is called an **irregular singularity** of the ODE. Let us start from the asymptotic form of the solutions for large x along the positive real axis, which are found to be

$$\psi(x) \sim cx^{-\frac{M}{2}} \exp\left(\pm \frac{1}{M+1} x^{M+1}\right), \quad x \to +\infty. \tag{1.2}$$

Here c is a constant. The solution with the plus sign shows a divergent behavior along the positive real axis while the solution corresponding to the minus sign decreases to zero. We call the solution the **subdominant solution** and denote it by $y(x, E)$. It is convenient to fix the normalization constant c in (1.2) of $y(x, E)$ as

$$y(x, E) \sim \frac{1}{\sqrt{2i}} x^{-\frac{M}{2}} \exp\left(-\frac{1}{M+1} x^{M+1}\right), \quad x \to +\infty. \tag{1.3}$$

The decaying property of the solution (1.3) holds as far as $\text{Re}(x^{M+1}) > 0$. Using the polar coordinates $x = re^{i\theta}$, this condition reads $\cos(M + 1)\theta > 0$, which means that

$$|\arg(x)| < \frac{\pi}{2M+2}. \tag{1.4}$$

The subdominant solution $y(x, E)$ is uniquely determined in this region. When $\arg(x)$ crosses the lines $\arg x = \pm \frac{\pi}{2M+2}$, $y(x, E)$ becomes the increasing solution. The subdominant solution in the remaining regions outside the region (1.4) can be constructed by **Sibuya's trick** [5] or the Symanzik rotation [16] explained as follows. By the scale transformation of x and E: $x \to ax$, $E \to a^{-2}E$, the ODE becomes

$$\left(-\frac{1}{a^{2M+2}} \frac{d^2}{dz^2} + z^{2M} - \frac{E}{a^{2M+2}}\right) \psi(az) = 0. \tag{1.5}$$

Then for a satisfying $a^{2M+2} = 1$, the ODE is invariant. Define ω as the $(2M + 2)$-th root of unity:

$$\omega := \exp\left(\frac{2\pi i}{2M+2}\right). \tag{1.6}$$

Fig. 1.1 Stokes sectors for
$2M = 3$

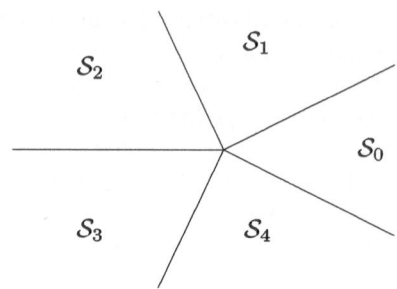

We define

$$y_k(x, E) := \omega^{\frac{k}{2}} y(\omega^{-k} x, \omega^2 E), \quad k \in \mathbb{Z}, \tag{1.7}$$

which is also the solution of (1.1). In particular, $y_0(x, E) = y(x, E)$. The solution $y_k(x, E)$ is subdominant in the region

$$S_k = \left\{ x; \; \left| \arg(x) - \frac{2\pi k}{2M + 2} \right| < \frac{\pi}{2M + 2} \right\}, \quad k \in \mathbb{Z}. \tag{1.8}$$

The complex plane is divided into the sectors S_k ($k \in \mathbb{Z}$) called the **Stokes sector**. The half-lines determined by $\arg(x) = \frac{(2k \pm 1)\pi}{2M+2}$, which are the boundaries of the Stokes sector S_k, are called the **anti-Stokes lines** (Fig. 1.1). When x crosses the anti-Stokes line, the sign of $\mathrm{Re}(x^{M+1})$ changes and the decaying solution (1.3) changes to the divergent one. The subdominant solution $y_k(x, E)$, which is uniquely defined in the Stokes sector S_k, is divergent in the neighboring sectors $S_{k\pm1}$.

We define the Wronskian of the functions f and g by

$$W[f, g] = f \frac{dg}{dx} - g \frac{df}{dx}. \tag{1.9}$$

It satisfies $W[f, g] = -W[g, f]$. For the solutions to the second-order ODE without the first derivative term, their Wrosnkian is independent of x. The Wronskian is zero if and only if the two solutions are linearly dependent. Since the Wronskian of y_k and y_{k+1} is non-zero, the pair $\{y_k, y_{k+1}\}$ provides a basis for the solutions of the ODE (1.1). We have fixed the normalization constant in the solution (1.3) such that

$$W[y_0, y_1] = 1. \tag{1.10}$$

For the solution (1.7), due to the factor $\omega^{\frac{k}{2}}$, $W[y_{k+1}, y_{k'+1}]$ and $W[y_k, y_{k'}]$ satisfy

$$W[y_{k+1}, y_{k'+1}](E) = W[y_k, y_{k'}](\omega^2 E), \quad k, k' \in \mathbb{Z}. \tag{1.11}$$

In particular, from $W[y_0, y_1] = 1$, the Wronskian of y_k, y_{k+1} is normalized as

$$W[y_k, y_{k+1}] = 1, \quad k \in \mathbb{Z}. \tag{1.12}$$

We now choose $\{y_0, y_1\}$ as the basis of the solutions. y_{-1} is expressed in terms of this basis as

$$y_{-1} = C(E)y_0 + \tilde{C}(E)y_1. \tag{1.13}$$

Taking the Wronskians of y_{-1} with y_0 and y_1, we obtain

$$\begin{aligned} W[y_0, y_{-1}] &= \tilde{C}(E)W[y_0, y_1], \\ W[y_1, y_{-1}] &= C(E)W[y_1, y_0]. \end{aligned} \tag{1.14}$$

From Eq. (1.12), we find

$$C(E) = W[y_{-1}, y_1], \quad \tilde{C}(E) = -1. \tag{1.15}$$

The coefficient $C(E)$ is called the **Stokes coefficient**, which relates the solutions with different asymptotics. When we choose another basis $\{y_{k-1}, y_k\}$, y_{-1} is expressed as

$$y_{-1} = C^{(k)}(E)y_{k-1} + \tilde{C}^{(k)}(E)y_k. \tag{1.16}$$

The Stokes coefficients can be expressed in terms of the Wronskian:

$$C^{(k)}(E) = W[y_{-1}, y_k], \quad \tilde{C}^{(k)}(E) = -W[y_{-1}, y_{k-1}]. \tag{1.17}$$

We now introduce the **T-function**:

$$T_k(E) = W[y_0, y_{k+1}](\omega^{-k-1}E), \quad k \in \mathbb{Z}. \tag{1.18}$$

In particular, we obtain $T_{-1}(E) = 0$, $T_0(E) = 1$. We also find

$$T_1(E) = W[y_0, y_2](\omega^{-2}E) = W[y_{-1}, y_1](E) = C(E). \tag{1.19}$$

In general, we obtain

$$\begin{aligned} C^{(k)}(E) &= W[y_0, y_{k+1}](\omega^{-2}E) = T_k(\omega^{k-1}E), \\ \tilde{C}^{(k)}(E) &= -W[y_0, y_k](\omega^{-2}E) = -T_{k-1}(\omega^{k-2}E). \end{aligned} \tag{1.20}$$

When $2M$ is a positive integer, the Stokes sector $\mathcal{S}_{2M+2}$ coincides with $\mathcal{S}_0$. y_{2M+2} and y_0 are linearly dependent. Due to the factor $\omega^{k/2}$ in (1.7), we find $y_{2M+2} = -y_0$ and $y_{2M+1} = -y_{-1}$. This implies that

$$T_{2M}(E) = 1, \quad T_{2M+1}(E) = 0. \tag{1.21}$$

The T-functions satisfy the functional relations which come from the identity satisfied by the determinants:

$$W[y_{k_1}, y_{k_2}]W[y_{k_3}, y_{k_4}] = W[y_{k_1}, y_{k_4}]W[y_{k_3}, y_{k_2}] + W[y_{k_3}, y_{k_1}]W[y_{k_4}, y_{k_2}]. \quad (1.22)$$

This relation is called the **Plücker identity**, which can be checked by direct computation. Now we choose $(k_1, k_2, k_3, k_4) = (1, k+2, 0, k+1)$. Then Eq. (1.22) reads

$$W[y_1, y_{k+2}]W[y_0, y_{k+1}] = W[y_1, y_{k+1}]W[y_0, y_{k+2}] + W[y_0, y_1]W[y_{k+1}, y_{k+2}]. \quad (1.23)$$

In terms of the T-functions, this relation reads

$$T_k(\omega^{k+3}E)T_k(\omega^{k+1}E) = T_{k-1}(\omega^{k+2}E)T_{k+1}(\omega^{k+2}E) + 1. \quad (1.24)$$

Changing the argument $E \to \omega^{-k-2}E$, we obtain

$$T_k(\omega E)T_k(\omega^{-1}E) = T_{k-1}(E)T_{k+1}(E) + 1, \quad k = 1, 2, \ldots. \quad (1.25)$$

This functional relation is called the **T-system**. For generic M, the relation includes an infinite number of the T-functions. When $2M$ is a positive integer, the T-functions satisfy the boundary conditions $T_0 = T_{2M} = 1$ and $T_{-1} = T_{2M+1} = 0$. Hence the T-system is truncated and only $T_k(E)$ $(k = 1, \ldots, 2M - 1)$ are nontrivial. The functional relations are called the A_{2M-1}-type T-system.

From the T-functions, we further define the **Y-function** $Y_k(E)$ by

$$Y_k(E) = T_{k-1}(E)T_{k+1}(E), \quad k \in \mathbb{Z}. \quad (1.26)$$

In particular, $Y_0(E) = 0$. From the T-system (1.25), the Y-functions are shown to satisfy the functional relations

$$Y_k(\omega E)Y_k(\omega^{-1}E) = \big(1 + Y_{k-1}(E)\big)\big(1 + Y_{k+1}(E)\big), \quad k = 1, 2, \ldots. \quad (1.27)$$

This functional relation is called the **Y-system**. For a positive integer $2M$, we also have the boundary condition: $Y_{2M}(E) = 0$, where $Y_1, \ldots, Y_{2M-1}$ are nontrivial. The relation (1.27) is called the A_{2M-1}-type Y-system. The Y-system plays a fundamental role in studying the thermodynamic properties of the integrable models which is explained in the Appendix. One can derive the integral equations for the Y-functions called the thermodynamic Bethe ansatz (TBA) equations and study the free energy of the system.

1.1.1 Numerical Test of the ODE/IM Correspondence

We will study the relation between the Stokes coefficients and the integrable models in detail. Let us write Eq. (1.13) using the relations (1.15) and (1.19) in the form

$$T_1(E)y_0(x, E) = y_1(x, E) + y_{-1}(x, E). \tag{1.28}$$

Since the ODE (1.1) does not have a singularity except at $x = \infty$, we can evaluate Eq. (1.28) at any finite point, say x_1:

$$T_1(E)y(x_1, E) = \omega^{\frac{1}{2}} y(x_1, \omega^2 E) + \omega^{-\frac{1}{2}} y(x_1, \omega^{-2} E). \tag{1.29}$$

This equation takes a similar form of **Baxter's T-Q relation** in the integrable models, where Q-function is identified as $y(x_1, E)$. The relation (1.29) provides the T-Q relation of the two-dimensional 6-vertex model or equivalently the one-dimensional quantum spin $1/2$ XXZ model, which is confirmed numerically as we will see below.

The one-dimensional spin $1/2$ XXZ model is defined on the lattice made of N points, where spin operators $\sigma_i^{x,y,z}$ ($i = 1, \ldots, N$) are assigned on each lattice point. The Hamiltonian is defined by

$$H_{XXZ} = -\frac{1}{2} \sum_{i=1}^{N} (\sigma_i^x \sigma_{i+1}^x + \sigma_i^y \sigma_{i+1}^y - \cos 2\eta \sigma_i^z \sigma_{i+1}^z), \tag{1.30}$$

together with the twisted boundary condition:

$$\sigma_1^z = \sigma_{N+1}^z, \quad (\sigma_1^x \pm i\sigma_1^y) = e^{i\phi}(\sigma_{N+1}^x \pm i\sigma_{N+1}^y). \tag{1.31}$$

The parameter η is called the anisotropy parameter and ϕ is the twisted parameter. When we take the conformal limit with $N \to \infty$, we can extract some functional relations from the model. First one can define the T-function $T(E)$ as the lowest eigenvalue of the transfer matrix which depends on the spectral parameter E. The transfer matrix is a generating function of conserved charges including the Hamiltonian H_{XXZ}. Baxter introduced the Q-functions to solve the eigenvalue problem of the spin chain model [8]. In the case of the XXZ spin chain, the T-Q relation for the T-function $T(E)$ and the Q-function $Q(E)$, which are the functions of the spectral parameter E, takes the form:

$$T(E)Q(E) = e^{i\phi} Q(\omega^2 E) + e^{-i\phi} Q(\omega^{-2} E), \tag{1.32}$$

where $\omega = \exp(i\pi - 2i\eta)$. Then (1.29) and (1.32) show the same relations if we identify $T_1(E)$ with $T(E)$, and $y(x_1, E)$ with $Q(E)$, together with the relations

$$\eta = \frac{\pi M}{2M + 2}, \quad \phi = \frac{\pi}{2M + 2}. \tag{1.33}$$

The function $Q(E)$ is an entire function in the complex plane with certain asymptotic behavior at large E [17]. The Hadamard factorization theorem tells us that $Q(E)$ is expressed in the form of the infinite product form $\prod_{n=1}^{\infty}(1 - \frac{E}{E_n})$. Evaluating (1.32) at a zero E_l ($l = 1, 2. \ldots$) of the Q-function, we obtain the set of equations called the **Bethe ansatz equations** (BAE):

$$\frac{Q(\omega^2 E_l)}{Q(\omega^{-2} E_l)} = -e^{2i\phi}, \quad l = 1, \ldots. \tag{1.34}$$

This can be written in the form:

$$\prod_{n=1}^{\infty} \frac{E_l - \omega^2 E_n}{E_l - \omega^{-2} E_n} = -e^{-2i\phi}, \quad l = 1, \ldots. \tag{1.35}$$

To find the Bethe roots E_l, the solutions of the BAE, it is useful to study the non-linear integral equation (NLIE) [12, 18] for the **counting function** $a(E) = e^{2i\phi}\frac{Q(\omega^2 E)}{Q(\omega^{-2} E)}$, where the BAE is equivalent to $a(E) = -1$.

For example, we calculate the Bethe roots E_n for the cubic potential x^3 with $M = \frac{3}{2}$ by solving the NLIE numerically. We find that the first two solutions to $a(E) = -1$ for $\ell = 0, -1$ are as in the following table:

$\ell = 0$	$\ell = -1$
3.45056275699435	1.02294784410254
9.52207607463254	6.37029280144269

Let us examine the Bethe roots from the viewpoint of the ODE. For the ODE (1.1) with $M = \frac{3}{2}$, let us consider the basis $\{\chi_0, \chi_1\}$ of the solutions around the origin, whose expansions are given by

$$\chi_0(x, E) = x - \frac{E}{6}x^3 + \frac{E^2}{120}x^5 + \frac{1}{30}x^6 + \cdots, \tag{1.36}$$

$$\chi_1(x, E) = 1 - \frac{E}{2}x^2 + \frac{E^2}{24}x^4 + \frac{1}{20}x^5 + \cdots. \tag{1.37}$$

Suppose that we can analytically continue $\chi_a(x, E)$ ($a = 1, 2$) to the subdominant solution $y(x, E)$ which is defined for large x, for a particular value of E. Then we expect that χ_a shows the decaying behavior at large x. One can estimate a value of E by looking for the solution $\chi_a(x_1, E) = 0$ at some fixed x_1. Let us fix $x_1 = 3.2$ for example. The first two solutions of $\chi_a(x_1, E) = 0$ for E on the real axis are found to be

$\chi_0(x_1, E) = 0$	$\chi_1(x_1, E) = 0$
3.4505911609160159133	1.0229497988456442809
9.5238133063575034068	6.3705568136520718095

The values of E surprisingly match each other. It looks strange that the correspondence depends on the choice of the value $x = x_1$. However, it turns out that there is little dependence on x_1 when x_1 is sufficiently large. The next section will present a refined version of the T-Q relation. It is interesting to explore why this correspondence holds, which is one of the main themes of this book. A further generalization of this observation will be seen in Chap. 3.

1.2 Schrödinger Equation with Centrifugal Potential Term

This section explores the more detailed correspondence between the second-order ordinary differential equations and the integrable model. Here we consider the Schrödinger equation with a centrifugal potential term

$$\left(-\frac{d^2}{dx^2} + \frac{\ell(\ell+1)}{x^2} + x^{2M} - E \right) \psi(x) = 0. \tag{1.38}$$

The new real parameter ℓ represents the angular momentum quantum number. Since the ODE is invariant under the symmetry $\ell \to -1 - \ell$, we consider the case where ℓ to $\ell \geq -\frac{1}{2}$. The ODE (1.38) has an irregular singularity at $x = \infty$ and $x = 0$ as a regular singularity.

Since the centrifugal term can be ignored at infinity, one can construct the asymptotic solutions at infinity as in the case of $\ell = 0$. The solution decaying along the positive real axis takes the same form as (1.3) which is denoted as $y(x, E, \ell)$. This is a solution that is subdominant in sector S_0, where the Stokes sectors S_k ($k \in \mathbb{Z}$) are defined by (1.8). Since the ODE is invariant under the rotation

$$(x, E, \ell) \to (\omega x, \omega^2 E, \ell), \quad \omega = e^{\frac{2\pi i}{2M+2}}, \tag{1.39}$$

the subdominant solution in the sector S_k is given by

$$y_k(x, E, \ell) := \omega^{\frac{k}{2}} y(\omega^{-k} x, \omega^{2k} E, \ell), \quad k \in \mathbb{Z}. \tag{1.40}$$

As in (1.28), $y_0, y_{\pm 1}$ satisfy the relation

$$T_1(E, \ell) y_0(x, E, \ell) = y_1(x, E, \ell) + y_{-1}(x, E, \ell), \tag{1.41}$$

where $T_1(E, \ell) = W[y_0, y_2](\omega^{-2}E, \ell)$.

Next, we consider the solution around $x = 0$, where the centrifugal potential is most relevant. One can construct two linearly independent power series solutions:

$$\chi^+(x, E, \ell) = x^{\ell+1} + \dots, \quad \chi^-(x, E, \ell) = x^{-\ell} + \dots, \tag{1.42}$$

where we considered the case $\ell \neq -\frac{1}{2}$. For $\ell = -\frac{1}{2}$, the solution includes the logarithm function of x. The Wronskian of χ_+ and χ_- is

$$W[\chi^+, \chi^-] = -(2\ell + 1), \tag{1.43}$$

which is non-zero for $\ell \neq -\frac{1}{2}$. We can perform the rotation (1.39), which gives the solution of the same ODE. For the basis $\chi_\pm(x, E, \ell)$, the rotation acts as

$$\chi_k^\pm(x, E, \ell) = \omega^{-k(\ell \pm \frac{1}{2})}\chi^\pm(x, E, \ell). \tag{1.44}$$

Note that $\chi^\pm(x, E, \ell)$ provide a diagonal basis under the monodromy transformation around $x = 0$:

$$\chi^\pm(e^{-2\pi i}x, E, \ell) = e^{\mp 2\pi i\ell}\chi^\pm(x, E, \ell). \tag{1.45}$$

Now we expand the subdominant solution $y_0(x, E, \ell)$ in terms of the basis $\{\chi^+, \chi^-\}$ as

$$(2\ell + 1)y_0(x, E, \ell) = -Q_-(E, \ell)\chi^+(x, E, \ell) + Q_+(E, \ell)\chi^-(x, E, \ell). \tag{1.46}$$

By taking the Wronskian with $\chi^\pm$, we find

$$Q_\pm(E, \ell) = W[y_0, \chi^\pm](E, \ell). \tag{1.47}$$

These two coefficients are related by $Q_+(E, \ell) = Q_-(E, -\ell - 1)$. Let us take the Wronskian of both sides (1.41) with $\chi^\pm$. We find

$$T_1(E, \ell)W[y_0, \chi^\pm] = W[y_{-1}, \chi^\pm] + W[y_1, \chi^\pm]. \tag{1.48}$$

Under rotation, one obtains

$$(2\ell + 1)y_k(x, E, \ell) = -\omega^{k(-\ell-\frac{1}{2})}Q_-(\omega^{2k}E, \ell)\chi^+(x, E, \ell)$$
$$+ \omega^{k(\ell+\frac{1}{2})}Q_+(\omega^{2k}E, \ell)\chi^-(x, E, \ell). \tag{1.49}$$

From

$$W[y_k(x, E, \ell), \chi_k^\pm(x, E, \ell)] = W[y_0, \chi^\pm](\omega^{2k}E, \ell), \tag{1.50}$$

and (1.44), we obtain

$$W[y_k, \chi^{\pm}](E, \ell) = \omega^{k(\ell \pm \frac{1}{2})} W[y_0, \chi^{\pm}](\omega^{2k} E, \ell). \tag{1.51}$$

Using Eq. (1.51) for $k = \pm 1$ and the definition (1.47), the relation (1.48) takes the form

$$T_1(E, \ell) Q_{\pm}(E, \ell) = \omega^{-(\ell \pm \frac{1}{2})} Q_{\pm}(\omega^{-2} E, \ell) + \omega^{(\ell \pm \frac{1}{2})} Q_{\pm}(\omega^2 E, \ell). \tag{1.52}$$

We have obtained again **Baxter's T-Q relation**, where $Q_{\pm}(E, \ell)$ is called the **Q-function**. Compared with (1.29), we have no x-coordinate dependence in (1.52), which provides a refined relation to the correspondence.

The Q-function $Q_{\pm}(E, \ell)$ is an entire function in the complex E-plane and has simple zeros at $E = E_a^{\pm}$ ($a = 0, 1, \ldots,$) [17]. Then evaluating the relation (1.52) at $E = E_a^{\pm}$ we obtain the Bethe ansatz equations:

$$\frac{Q_{\pm}(\omega^2 E, \ell)}{Q_{\pm}(\omega^{-2} E, \ell)} \bigg|_{E=E_a^{\pm}} = -\omega^{-2(\ell \pm \frac{1}{2})}. \tag{1.53}$$

Now we can map the parameters M and ℓ in the ODE side to those of the spin $1/2$ XXZ spin chain by

$$\eta = \frac{\pi M}{2M + 2}, \quad \phi = \frac{\pi (2\ell + 1)}{2M + 2}, \tag{1.54}$$

where the twisted parameter is related to the angular momentum ℓ. In the conformal limit of the six-vertex model with a twisted parameter ϕ, the theory is described by conformal field theory (CFT), whose effective central charge is given by [11]

$$c_{\text{eff}} = 1 - \frac{6\phi^2}{\pi(\pi - 2\eta)}. \tag{1.55}$$

In a series of papers by Bazhanov, Lukyanov, and Zamolodchikov [19–22], the Q-functions and the T-functions of the spin-1/2 XXZ model are constructed from the conformal field theory deformed by a holomorphic primary field. The CFT is realized by a free field and has a central charge

$$c = 1 - 6 \left(\beta - \frac{1}{\beta} \right)^2. \tag{1.56}$$

The primary field is a vertex operator with momentum p which has the conformal dimension

$$\Delta = \left(\frac{p}{\beta} \right)^2 + \frac{c - 1}{24}. \tag{1.57}$$

The Q-functions and T-functions in CFT perturbed by the operator with momentum p satisfy the T-Q relation with

$$\beta^2 = 1 - \frac{2\eta}{\pi}, \quad p = \frac{\phi}{2\pi}. \tag{1.58}$$

Combining the two parameter identifications (1.54) and (1.58), we find that the central charge for the ODE (1.38) is

$$c = 1 - \frac{6M^2}{M+1}. \tag{1.59}$$

This CFT belongs to the non-unitary minimal series $M_{2,2M+2}$ when $2M$ is an odd integer. In particular, for $2M = 3$, the minimal model $M_{2,5}$ is called the Yang–Lee edge singularity, which has the central charge $c = -\frac{22}{5}$. The CFT contains two primary fields, the identity operator and $\phi_{1,3}$ with $\Delta = -\frac{1}{5}$. The effective central charge is

$$c_{\text{eff}} := c - 24\Delta = 1 - \frac{6(\ell + \frac{1}{2})^2}{M+1}, \tag{1.60}$$

which is equal to (1.55) under the relations (1.58). Some basic notions of CFT and further numerical evidence will be explained in the Appendix.

1.2.1 T-System and Y-System

The T- and Y-systems presented in Sect. 1.1 are also modified by the singularity at the origin. Here we will consider the case where $2M$ is a positive integer. The basis $\{y_0, y_1\}$ will change non-trivially under the monodromy transformation around $x = 0$, which is represented as

$$\begin{pmatrix} y_1 \\ y_0 \end{pmatrix} (e^{-2\pi i}x, E, \ell) = \Omega(E, \ell) \begin{pmatrix} y_1 \\ y_0 \end{pmatrix} (x, E, \ell), \tag{1.61}$$

where $\Omega(E, \ell)$ is a 2×2 matrix with non-diagonal elements in general. For $\ell = 0, -1$, since there is no singularity in the complex plane, Ω is proportional to the identity matrix. In the basis $\{\chi^+, \chi^-\}$, the monodromy matrix $\mathcal{M}(E, \ell)$ becomes diagonal:

$$\begin{pmatrix} \chi^+ \\ \chi^- \end{pmatrix} (e^{-2\pi i}x, E, \ell) = \mathcal{M}(E, \ell) \begin{pmatrix} \chi^+ \\ \chi^- \end{pmatrix} (x, E, \ell), \quad \mathcal{M}(E, \ell) = \begin{pmatrix} e^{2\pi i\ell} & 0 \\ 0 & e^{-2\pi i\ell} \end{pmatrix}. \tag{1.62}$$

The matrices Ω and $\mathcal{M}$ are related to each other by the change of basis (1.46):

$$\Omega = Q\mathcal{M}Q^{-1}, \quad \begin{pmatrix} y_1 \\ y_0 \end{pmatrix} = Q \begin{pmatrix} \chi^+ \\ \chi^- \end{pmatrix}, \tag{1.63}$$

where Q is a 2×2 matrix. Note that the trace of a monodromy matrix is invariant and is evaluated as

$$\mathrm{tr}\,\Omega = \mathrm{tr}\,\mathcal{M} = 2\cos(2\pi\ell). \tag{1.64}$$

We will express the monodromy matrix in terms of the Wronskians. The Stokes sector S_{2M+2+j} completely overlaps with S_j for an integer $2M$. Then, we obtain

$$y_{2M+2+j}(x, E, \ell) = \omega^{\frac{2M+2}{2}} y_j(e^{-2\pi i}x, E, \ell), \quad j = 0, 1. \tag{1.65}$$

In terms of the basis $\{y_0, y_1\}$, y_{2M+2+j} is expanded as

$$y_{2M+2+j}(x, E, \ell) = -W[y_1, y_{2M+2+j}]y_0(x, E, \ell) + W[y_0, y_{2M+2+j}]y_1(x, E, \ell). \tag{1.66}$$

Combining (1.65) and (1.66), we find

$$\Omega(E, \ell) = \omega^{-(M+1)} \begin{pmatrix} W[y_0, y_{2M+3}] & -W[y_1, y_{2M+3}] \\ W[y_0, y_{2M+2}] & -W[y_1, y_{2M+2}] \end{pmatrix}. \tag{1.67}$$

In particular, one obtains

$$\mathrm{tr}\,\Omega = \omega^{-(M+1)}(W[y_0, y_{2M+3}] - W[y_1, y_{2M+2}]). \tag{1.68}$$

Now we introduce the T-functions as in (1.18):

$$T_a(E, \ell) = W[y_0, y_{a+1}](\omega^{-a-1}E, \ell), \quad a \in \mathbb{Z}. \tag{1.69}$$

Then the T-functions (1.69) satisfy the T-system (1.25), which comes from the Plücker identities. The boundary condition at $a = -1, 0$ is given by $T_{-1} = 0$ and $T_0 = 1$. At $a = 2M + 1$, the condition differs from (1.21). In fact, Eq. (1.68) is written as

$$\mathrm{tr}\,\Omega = \omega^{-(M+1)} \left(T_{2M+2}(\omega^{2M+3}E, \ell) - T_{2M}(\omega^{2M+3}E, \ell) \right). \tag{1.70}$$

The T-system does not truncate and continues to infinity.

Now we define the Y-functions as in (1.26):

$$Y_a(E, \ell) = T_{a-1}(E, \ell)T_{a+1}(E, \ell), \quad a = 0, 1, 2, \ldots, \tag{1.71}$$

which satisfy the Y-system (1.27). Here we define $Y_0 = 0$. For $a = 2M + 1$, we have

$$Y_{2M+1}(E, \ell) = T_{2M}(E, \ell)T_{2M+2}(E, \ell). \tag{1.72}$$

Multiplying Eq. (1.70) by $T_{2M}(\omega^2 E, \ell)$ and replacing $E \rightarrow \omega^{-2M-3}E$, we obtain

$$Y_{2M+1}(E, \ell) = 2\omega^{M+1} \cos(2\pi\ell)T_{2M}(E, \ell) + T_{2M}(E, \ell)^2, \tag{1.73}$$

where we have used (1.64). We add the new Y-function:

$$\hat{Y}(E, \ell) := T_{2M}(E, \ell). \tag{1.74}$$

Then $Y_a(E, \ell)$ $(a = 1, \ldots, 2M)$ and $\hat{Y}(E, \ell)$ form a closed relation:

$$Y_a(\omega E, \ell)Y_a(\omega^{-1}E, \ell) = (1 + Y_{a-1}(E, \ell))(1 + Y_{a+1}(E, \ell)), \quad a = 1, \ldots, 2M - 1, \tag{1.75}$$

$$Y_{2M}(\omega E, \ell)Y_{2M}(\omega^{-1}E, \ell) = (1 + Y_{2M-1}(E, \ell))$$
$$\times (1 + \omega^{\frac{M+1}{2}} e^{2\pi i\ell}\hat{Y}(E, \ell))(1 + \omega^{\frac{M+1}{2}} e^{-2\pi i\ell}\hat{Y}(E, \ell)), \tag{1.76}$$

$$\hat{Y}(\omega E, \ell)\hat{Y}(\omega^{-1}E, \ell) = 1 + Y_{2M}(E, \ell). \tag{1.77}$$

This is the Y-system of D_{2M+2}-type. We can convert this Y-system into a set of TBA equations. The effective central charge becomes (1.60) [23].

To summarize, we present in Table 1.1 the ODE/IM correspondence. The Y-functions are obtained from the T-functions. It is not obvious, at present, what kind of object on the ODE side corresponds to the Y-functions and also the meaning of the TBA equations. This will become clear in the following chapters. For $\ell = 0, -1$, we have seen $y_{2M+2} = -y_0$ and $y_{2M+3} = -y_1$, which means that Ω is the minus of the identity matrix. In this case, $T_{2M+1} = 0$ such that $Y_{2M} = 0$. We thus obtain the truncated Y-system

$$Y_a(\theta + \frac{\pi}{2M})Y_a(\theta - \frac{\pi}{2M}) = (1 + Y_{a+1}(\theta))(1 + Y_{a-1}(\theta)), \quad a = 1, \ldots, 2M - 1, \tag{1.78}$$

Table 1.1 Correspondence between the Schrödinger equation and the XXZ model

ODE	IM
Order of the potential $2M$	Anisotropy parameter η
Angular momentum ℓ	Twist parameter ϕ
Energy E	Spectral parameter θ

where we have introduced the spectral parameter θ by $E = \exp(\theta/\mu)$ with $\mu = \frac{M+1}{2M}$.

We further explore the correspondence by rewriting the Y-system into the integral equation. To derive the integral equations, we first evaluate the asymptotic behaviors of the Y-functions for large and small E. At large E, which corresponds to large θ, the Y-functions behave as

$$\log Y_a(\theta) \sim m_a L e^{\theta}, \quad |E| \to \infty, \quad |\mathrm{Arg}(E)| < \pi. \tag{1.79}$$

Substituting this asymptotic behavior into the Y-system, the masses m_a must satisfy the constraints

$$2\cos\frac{\pi}{2M} = \frac{m_{a-1} + m_{a+1}}{m_a}, \tag{1.80}$$

which can be solved as

$$m_a L = c \sin\frac{\pi a}{2M}. \tag{1.81}$$

Here c is a constant, which is determined as follows. To fix the constant value c, we first compute the Q-function at large E. Since the Wronskian is independent of the coordinate z, we can evaluate the Q-function (1.47) at $z \to 0$:

$$Q_+(E) = \lim_{z\to 0} W[y_0, \psi_+](E) = \lim_{z\to 0}\left((\ell+1)z^{\ell}y(z, E)\right). \tag{1.82}$$

Substituting the WKB approximation of y into (1.82), one finds the behavior of the Q-function for large E [7, 17]:

$$\log Q_+(E) \sim \frac{a_0}{2}(-E)^{\mu}, \quad |E| \to \infty \text{ and } |\arg(-E)| < \pi, \tag{1.83}$$

where constant a_0 is given by

$$\frac{a_0}{2} = \int_0^{\infty} dt\left(\sqrt{t^{2M}-1} - t^{2M}\right) = \frac{\Gamma(1+\frac{1}{2M})\Gamma(\frac{3}{2})}{\Gamma(\frac{3}{2}+\frac{1}{2M})\sin(\frac{\pi}{2}+\frac{\pi}{2M})}. \tag{1.84}$$

We then evaluate the large E behavior of T-functions. Substituting the expansion of y_k (1.49) into the T-function T_M, one finds

$$(2\ell+1)T_M(E) = \omega^{(M+1(\ell+\frac{1}{2})}Q_+(\omega^{M+1}E)Q_-(\omega^{-(M+1)}E) \\ - \omega^{-(M+1)(\ell+\frac{1}{2})}Q_+(\omega^{-(M+1)}E)Q_-(\omega^{(M+1)}E). \tag{1.85}$$

For $\ell = 0$, this equation reduces to[1]

$$T_M(E) = 2i\, Q_+(-E)Q_-(-E). \tag{1.86}$$

---

[1] This T-function T_M is shown to be the spectral determinant $D(E)$ in [24].

We thus find the asymptotic behavior of $T_M(-E)$:

$$\log T_M(-E) \sim \log Q_+(E)Q_-(E) = a_0(-E)^\mu, \quad |E| \to \infty, \quad |\arg(E)| < \pi. \tag{1.87}$$

Recall the T-system (1.25) for $k = M$ reads

$$T_M(\omega^{-1}E)T_M(\omega E) = 1 + Y_M(E). \tag{1.88}$$

Then, the asymptotic behavior of the Y-function $Y_M(-E)$ for large E is given by

$$\log Y_M(-E) \sim \log T_M(-\omega^{-1}E) + \log T_M(-\omega E) \sim 2\cos\frac{\pi}{2M}a_0(-E)^\mu, \quad |E| \to \infty. \tag{1.89}$$

The constant c in (1.81) is thus fixed as

$$c = 2a_0\cos\frac{\pi}{2M}. \tag{1.90}$$

In the limit $\theta \to -\infty$, which corresponds to $E \to 0$, Y_a goes to the constant $Y_a \to Y_a^*$, which is given by [25]

$$Y_a^* = \frac{\sin(\frac{a\pi}{2M+1})\sin(\frac{(a+2)\pi}{2M+1})}{\sin^2(\frac{\pi}{2M+1})}. \tag{1.91}$$

Now we convert the Y-system into the integral equations. Introducing $\epsilon_a(\theta) = \log Y_a(\theta)$ and $L_a(\theta) = \log(1 + e^{-\epsilon_a(\theta)})$, (1.78) becomes

$$\epsilon_a(\theta + \frac{i\pi}{2M}) + \epsilon_a(\theta - \frac{i\pi}{2M}) - \sum_{b=1}^{2M-1} I_{ab}\epsilon_b(\theta) = \sum_{b=1}^{2M-1} I_{ab}L_b(\theta). \tag{1.92}$$

Here $I_{ab} = \delta_{a-1,b} + \delta_{a+1,b}$, which is regarded as the incidence matrix of the Lie algebra A_{2M-1}. The constraints (1.80) for m_a can be written as

$$2\cosh\frac{\pi a}{2M}m_a = \sum_l I_{ab}m_b. \tag{1.93}$$

We can replace $\epsilon_a(\theta)$ to $f_a(\theta) := \epsilon_a(\theta) - m_a e^\theta$ in (1.92). We then perform the Fourier transformation of (1.92), which takes the form

$$\sum_{b=1}^{2M-1} (2\cosh\frac{\pi k}{2M}\delta_{ab} - I_{ab})\tilde{f}_b(k) = \sum_{b=1}^{2M-1} \tilde{L}_b(\theta), \tag{1.94}$$

where we define

$$f(\theta) = \int_{-\infty}^{\infty} e^{ik\theta} \tilde{f}(k) dk. \tag{1.95}$$

We solve (1.94) for $\tilde{\epsilon}_a(k)$ and apply the inverse Fourier transformation to find the equation for $\epsilon_a(theta)$, which is found to be

$$\epsilon_a(\theta) = m_a e^{\theta} + \sum_{b=1}^{2M-1} \phi_{ab} * L_b(\theta), \tag{1.96}$$

where $f * g(\theta)$ denotes the convolution of the functions $f(\theta)$ and $g(\theta)$:

$$(f * g)(\theta) = \int_{-\infty}^{\infty} d\theta' f(\theta - \theta') g(\theta'). \tag{1.97}$$

$\phi_{ab}(\theta)$ is defined by

$$\phi_{ab}(\theta) = \int dk e^{ik\theta} \tilde{\phi}_{ab}(k), \quad \tilde{\phi}_{ab}(k) = \left(2 \cosh \frac{\pi k}{2M} \delta_{ac} - I_{ac}\right)^{-1} I_{cb}. \tag{1.98}$$

The integral equation (1.96) is called the **TBA (thermodynamic Bethe ansatz) equation**. The TBA equations determine the energies $\epsilon_a(\theta)$ of the pseudo-particles labeled by a in a two-dimensional integrable field theory. For Eq. (1.96), the system corresponds to the conformal or kink limit of the massive integrable field theory associated with $A_{2M-1}^{(1)}$-type S-matrix as seen in the Appendix. The effective central charge of the TBA system is defined by

$$c_{\text{eff}} = \frac{3}{2\pi^2} \sum_{a=1}^{2M-1} \int_{-\infty}^{\infty} m_a e^{\theta} L_a(\theta). \tag{1.99}$$

Using the dilogarithm identity [26], we find

$$c_{\text{eff}} = \frac{2M-1}{2M+1}. \tag{1.100}$$

This is equal to half the central charge of the $\mathbb{Z}_{2M}$ parafermion theory. The TBA equation for $\ell \neq 0, -1$ can also be derived in a similar way [23]. Using the relation (1.33), we find that the value of effective central charge (1.55) agrees with the TBA result.

Table 1.2 The ODE/IM correspondence

ODE	IM
Energy	Spectral parameter
Connection coefficients between 0 and ∞	Q-function (T-Q relation)
Stokes coefficients = Wronskians	T-functions
Plücker relation	T-system

1.3 Summary and Outlook

To summarize, we present the following correspondence between the ordinary differential equations and the quantum integrable models (Table 1.2).

Here we will make some comments related to the topics in this chapter. In [20], the Q-operator has been constructed from CFT, where the Q-operator acts on the highest weight Virasoro module V_Δ with a conformal weight Δ given by Eq. (1.57). In the work of Dorey–Tateo [17], the spectral determinant of the Schrödinger equation (1.1) was observed to coincide with the eigenfunction of the Q-operator with the special value of the momentum p.[2] This is generalized to the Q-operators of the Virasoro module for general p [9], which coincide with the Schrödinger equation (1.38).[3] This correspondence is generalized to the states with higher levels in the Virasoro module [28], where the corresponding Schrödinger equation has the "monster potential",

$$V(x) = x^{2M} + \frac{\ell(\ell+1)}{x^2} - 2\frac{d^2}{dx^2}\sum_{k=1}^{L}\log(x^{2M+2} - z_k). \tag{1.101}$$

See [29–36] for further developments in this direction.

The ODE/IM correspondence for the Schrödinger equation with inhomogeneous potential

$$\left(-\frac{d^2}{dx^2} + x^{2M} + \alpha x^{M-1} + \frac{\ell(\ell+1)}{x^2} - E\right)\psi(x) = 0 \tag{1.102}$$

has been explored in [37, 38], which corresponds to the $U_q(\widehat{gl}(2|1))$ quantum integrable model with coupled BAEs/NLIEs. Interestingly, this type of ODE is related to the PT-symmetric Schrödinger equation, where the reality of the energy spectrum has been proved based on the corresponding BAEs [38–42].

The Schrödinger with non-simple turning points, $p(x) = (x^{2M/k} - E)^k$, have been studied in [43, 44]. In the next chapter, we will expand the ODE/IM correspondence to the Schrödinger equation with an arbitrary polynomial potential. Other interesting generalizations of ODE/IM correspondence for the Schrödinger equation

---

[2] See also the Appendix for the integrability structure of the Virasoro module for $c < 1$ CFT.

[3] See also [27] for this correspondence.

includes the Heun-type equation [45], the paperclip models [46, 47], the Mathieu equation and its modification [48–53], and the Kondo problems [54–56].

The ODE/IM correspondence has been generalized to the higher-order ODEs with a homogeneous potential. In the case of an ODE that has no singularity at the origin [57–59]

$$\left((-1)^{r+2} \frac{d^{r+1}}{dx^{r+1}} + P(x) \right) \psi(x) = 0, \quad P(x) = x^{(r+1)M} - E, M > 0, \quad (1.103)$$

the Y-system has been obtained from the Wronskian.[4] From the exact WKB analysis developed in the next chapter, we may calculate the asymptotics of the Y-function at large E, which leads to a set of TBA equations. For the quadratic potential, $P(x) = x^2 - E$, the TBA equations have been derived in [60]. Interestingly, the resulting TBA equations have the same form as those in Eq. (1.96). For a general value of $(r + 1)M$, the corresponding quantum integrable model for the ODE (1.103) has been studied based on the Bethe ansatz equation and NLIE approach [27, 43, 61], which is proposed to be

$$\frac{(A_r)_L \times (A_r)_1}{(A_r)_{L+1}} \quad (1.104)$$

with $L = \frac{1}{M} - (r + 1)$. The effective central charge of this non-unitary CFT is

$$c_{\text{eff}} = \frac{r\big((r + 1)M - 1\big)}{(r + 1)(M + 1)}. \quad (1.105)$$

When $r = 1$ and $2M = m + 1$, $c_{\text{eff}} = \frac{m}{m+3}$, which is the same as the case $r = m$ and $(r + 1)M = 2$. Also from our observation of the coincidence of the TBA equations, these two types of ODE correspond to the same quantum integrable model.

The higher-order ODE of the type (1.103) is further generalized by introducing the angular momentum part and including the pseudo-differential operator in [10, 43, 61, 62], which correspond to the Bethe ansatz equations for the classical Lie algebras $\mathfrak{g}$. We will return to this higher-order ODE from a different viewpoint at the end of this book.

References

1. G. Wentzel, Eine Verallgemeinerung der Quantenbedingungen für die Zwecke der Wellen-mechanik. Z. Physik **38**, 518–529 (1926). https://doi.org/10.1007/BF01397171
2. L. Brillouin, La mecanique ondulatoire de Schrödinger; une méthode générale de resolution par approximations successives. Comptes Rendus **183**, 24–26 (1926)
3. H. Kramers, Zeits. f. Phyzik **39**, 828 (1926)

---

[4] See Chap. 3 for the massive version of this Y-system.

4. A. Voros, The return of the quartic oscillator. The complex WKB method. Annales de l'I.H.P. Physique théorique **39**(3), 211–338 (1983)

5. Y. Sibuya, *Global Theory of a Second Order Linear Ordinary Differential Equation with a Polynomial Coefficient* (North-Holland, 1975). North-Holland Mathematics Studies, Vol. 18

6. P. Dorey, R. Tateo, Anharmonic oscillators, the thermodynamic Bethe ansatz, and nonlinear integral equations. J. Phys. **A32**, L419–L425 (1999). https://doi.org/10.1088/0305-4470/32/38/102, arXiv:hep-th/9812211 [hep-th]

7. P. Dorey, C. Dunning, R. Tateo, The ODE/IM Correspondence. J. Phys. A **40**, R205 (2007). https://doi.org/10.1088/1751-8113/40/32/R01, arXiv:hep-th/0703066

8. R.J. Baxter, Partition function of the eight vertex lattice model. Annals Phys. **70**, 193–228 (1972). https://doi.org/10.1016/0003-4916(72)90335-1

9. V.V. Bazhanov, S.L. Lukyanov, A.B. Zamolodchikov, Spectral determinants for Schrodinger equation and Q operators of conformal field theory. J. Statist. Phys. **102**, 567–576 (2001). https://doi.org/10.1023/A:1004838616921, arXiv:hep-th/9812247

10. P. Dorey, R. Tateo, Differential equations and integrable models: The SU(3) case. Nucl. Phys. **B571**, 583–606 (2000). https://doi.org/10.1016/S0550-3213(99)00791-9, https://doi.org/10.1016/S0550-3213(01)00164-X, arXiv:hep-th/9910102 [hep-th]. [Erratum: Nucl. Phys.B603,582(2001)]

11. A. Klumper, M.T. Batchelor, P.A. Pearce, Central charges of the 6- and 19- vertex models with twisted boundary conditions. J. Phys. A **24**, 3111 (1991)

12. C. Destri, H.J. de Vega, New thermodynamic Bethe ansatz equations without strings. Phys. Rev. Lett. **69**, 2313–2317 (1992). https://doi.org/10.1103/PhysRevLett.69.2313

13. A.B. Zamolodchikov, On the thermodynamic Bethe ansatz equations for reflectionless ADE scattering theories. Phys. Lett. **B253**, 391–394 (1991). https://doi.org/10.1016/0370-2693(91)91737-G

14. A. Kuniba, T. Nakanishi, J. Suzuki, Functional relations in solvable lattice models. 1: Functional relations and representation theory. Int. J. Mod. Phys. A **9**, 5215–5266 (1994). https://doi.org/10.1142/S0217751X94002119. arXiv:hep-th/9309137

15. A. Kuniba, T. Nakanishi, J. Suzuki, T-systems and Y-systems in integrable systems. J. Phys. A **44**, 103001 (2011). https://doi.org/10.1088/1751-8113/44/10/103001. arXiv:1010.1344 [hep-th]

16. A. Voros, From exact-WKB towards singular quantum perturbation theory. Publ. Res. Inst. Math. Sci. **40**(3), 973–990 (2004). https://doi.org/10.2977/PRIMS/1145475499

17. P. Dorey, R. Tateo, On the relation between Stokes multipliers and the T-Q systems of conformal field theory. Nucl. Phys. B **563**, 573–602 (1999). https://doi.org/10.1016/S0550-3213(99)00609-4, arXiv:hep-th/9906219. [Erratum: Nucl. Phys. B 603, 581–581 (2001)]

18. A. Klümper, P.A. Pearce, Analytic calculation of scaling dimensions: Tricritical hard squares and critical hard hexagons. J. Stat. Phys. **64**(1), 13–76 (1991). https://doi.org/10.1007/BF01057867

19. V.V. Bazhanov, S.L. Lukyanov, A.B. Zamolodchikov, Integrable structure of conformal field theory, quantum KdV theory and thermodynamic Bethe ansatz. Commun. Math. Phys. **177**, 381–398 (1996). https://doi.org/10.1007/BF02101898, arXiv:hep-th/9412229

20. V.V. Bazhanov, S.L. Lukyanov, A.B. Zamolodchikov, Integrable structure of conformal field theory. 2. Q operator and DDV equation. Commun. Math. Phys. **190**, 247–278 (1997). https://doi.org/10.1007/s002200050240, arXiv:hep-th/9604044

21. V.V. Bazhanov, S.L. Lukyanov, A.B. Zamolodchikov, Integrable quantum field theories in finite volume: Excited state energies. Nucl. Phys. B **489**, 487–531 (1997). https://doi.org/10.1016/S0550-3213(97)00022-9, arXiv:hep-th/9607099

22. V.V. Bazhanov, S.L. Lukyanov, A.B. Zamolodchikov, Integrable structure of conformal field theory. 3. The Yang-Baxter relation. Commun. Math. Phys. **200**, 297–324 (1999). https://doi.org/10.1007/s002200050531, arXiv:hep-th/9805008

23. K. Ito, H. Shu, TBA equations for the Schrödinger equation with a regular singularity. J. Phys. A **53**(33), 335201 (2020). https://doi.org/10.1088/1751-8121/ab96ee, arXiv:1910.09406 [hep-th]

24. J. Suzuki, Anharmonic oscillators, spectral determinant and short exact sequence of U(q) (affine sl(2)). J. Phys. A **32**, L183–L188 (1999). https://doi.org/10.1088/0305-4470/32/16/002, arXiv:hep-th/9902053

25. N. Yu. Reshetikhin, P.B. Wiegmann, Towards the classification of completely integrable quantum field theories. Phys. Lett. **B189**, 125–131 (1987). https://doi.org/10.1016/0370-2693(87)91282-2

26. A.N. Kirillov, Identities for the Rogers dilogarithm function connected with simple Lie algebras. J. Soviet Mathem. **47**, 2450–2459 (1989). https://doi.org/10.1007/BF01840426

27. P. Dorey, C. Dunning, F. Gliozzi, R. Tateo, On the ODE/IM correspondence for minimal models. J. Phys. A **41**, 132001 (2008). https://doi.org/10.1088/1751-8113/41/13/132001, arXiv:0712.2010 [hep-th]

28. V.V. Bazhanov, S.L. Lukyanov, A.B. Zamolodchikov, Higher level eigenvalues of Q operators and Schroedinger equation. Adv. Theor. Math. Phys. **7**(4), 711–725 (2003). https://doi.org/10.4310/ATMP.2003.v7.n4.a4, arXiv:hep-th/0307108

29. D. Fioravanti, F. Ravanini, M. Stanishkov, Generalized KdV and quantum inverse scattering description of conformal minimal models. Phys. Lett. B **367**, 113–120 (1996). https://doi.org/10.1016/0370-2693(95)01463-2, arXiv:hep-th/9510047

30. B. Feigin, E. Frenkel, Quantization of soliton systems and Langlands duality. arXiv:0705.2486 [math.QA]

31. E. Frenkel, D. Hernandez, Spectra of quantum KdV Hamiltonians, Langlands duality, and affine opers. Commun. Math. Phys. **362**(2), 361–414 (2018). https://doi.org/10.1007/s00220-018-3194-9, arXiv:1606.05301 [math.QA]

32. E. Frenkel, P. Koroteev, D.S. Sage, A.M. Zeitlin, q-Opers, QQ-Systems, and Bethe Ansatz. arXiv:2002.07344 [math.AG]

33. D. Masoero, A. Raimondo, Opers for higher states of quantum KdV models. Commun. Math. Phys. **378**(1), 1–74 (2020). https://doi.org/10.1007/s00220-020-03792-3, arXiv:1812.00228 [math-ph]

34. R. Conti, D. Masoero, Counting monster potentials. JHEP **02**, 059 (2021). https://doi.org/10.1007/JHEP02(2021)059, arXiv:2009.14638 [math-ph]

35. R. Conti, D. Masoero, On solutions of the Bethe Ansatz for the Quantum KdV model. arXiv:2112.14625 [math-ph]

36. D. Fioravanti, M. Rossi, On the origin of the correspondence between classical and quantum integrable theories. Phys. Lett. B **838**, 137706 (2023). https://doi.org/10.1016/j.physletb.2023.137706, arXiv:2106.07600 [hep-th]

37. J. Suzuki, Functional relations in Stokes multipliers: Fun with x**6 + alpha x**2 potential. J. Statist. Phys. **102**, 1029–1047 (2001). https://doi.org/10.1023/A:1004823608260, arXiv:quant-ph/0003066

38. P. Dorey, C. Dunning, R. Tateo, Spectral equivalences, Bethe Ansatz equations, and reality properties in PT-symmetric quantum mechanics. J. Phys. A **34**, 5679–5704 (2001). https://doi.org/10.1088/0305-4470/34/28/305, arXiv:hep-th/0103051

39. P. Dorey, C. Dunning, R. Tateo, Supersymmetry and the spontaneous breakdown of PT symmetry. J. Phys. A **34**, L391 (2001). https://doi.org/10.1088/0305-4470/34/28/102, arXiv:hep-th/0104119

40. P. Dorey, C. Dunning, R. Tateo, A Reality proof in PT symmetric quantum mechanics. Czech. J. Phys. **54**, 35–41 (2004). https://doi.org/10.1023/B:CJOP.0000014365.19507.b6, arXiv:hep-th/0309209

41. P. Dorey, A. Millican-Slater, R. Tateo, Beyond the WKB approximation in PT-symmetric quantum mechanics. J. Phys. A **38**, 1305–1332 (2005). https://doi.org/10.1088/0305-4470/38/6/010, arXiv:hep-th/0410013

42. P. Dorey, C. Dunning, A. Lishman, R. Tateo, PT symmetry breaking and exceptional points for a class of inhomogeneous complex potentials. J. Phys. A **42**, 465302 (2009). https://doi.org/10.1088/1751-8113/42/46/465302, arXiv:0907.3673 [hep-th]

43. P. Dorey, C. Dunning, D. Masoero, J. Suzuki, R. Tateo, Pseudo-differential equations, and the Bethe ansatz for the classical Lie algebras. Nucl. Phys. **B772**, 249–289 (2007). https://doi.org/10.1016/j.nuclphysb.2007.02.029, arXiv:hep-th/0612298 [hep-th]

44. J. Suzuki, Elementary functions in Thermodynamic Bethe Ansatz. J. Phys. A **48**(20), 205204 (2015). https://doi.org/10.1088/1751-8113/48/20/205204, arXiv:1501.00773 [math-ph]

45. P. Dorey, J. Suzuki, R. Tateo, Finite lattice Bethe ansatz systems and the Heun equation. J. Phys. A **37**, 2047–2062 (2004). https://doi.org/10.1088/0305-4470/37/6/006, arXiv:hep-th/0308053

46. S.L. Lukyanov, E.S. Vitchev, A.B. Zamolodchikov, Integrable model of boundary interaction: The Paperclip. Nucl. Phys. B **683**, 423–454 (2004). https://doi.org/10.1016/j.nuclphysb.2004.02.010, arXiv:hep-th/0312168

47. S.L. Lukyanov, A.M. Tsvelik, A.B. Zamolodchikov, Paperclip at theta = pi. Nucl. Phys. B **719**, 103–120 (2005). https://doi.org/10.1016/j.nuclphysb.2005.04.040, arXiv:hep-th/0501155

48. A.B. Zamolodchikov, Generalized Mathieu equations and Liouville TBA, in *Quantum Field Theories in Two Dimensions*, vol. 2. World Scientific (2012)

49. S.L. Lukyanov, A.B. Zamolodchikov, Integrable circular brane model and Coulomb charging at large conduction. J. Stat. Mech. **0405**, P05003 (2004). https://doi.org/10.1088/1742-5468/2004/05/P05003, arXiv:hep-th/0306188

50. V.A. Fateev, S.L. Lukyanov, Boundary RG flow associated with the AKNS soliton hierarchy. J. Phys. A **39**, 12889–12926 (2006). https://doi.org/10.1088/0305-4470/39/41/S10, arXiv:hep-th/0510271

51. S.L. Lukyanov, Notes on parafermionic QFT's with boundary interaction. Nucl. Phys. B **784**, 151–201 (2007). https://doi.org/10.1016/j.nuclphysb.2007.04.034, arXiv:hep-th/0606155

52. D. Fioravanti, D. Gregori, Integrability and cycles of deformed N = 2 gauge theory. Phys. Lett. B **804**, 135376 (2020). https://doi.org/10.1016/j.physletb.2020.135376, arXiv:1908.08030 [hep-th]

53. D. Fioravanti, D. Gregori, H. Shu, Integrability, susy $SU(2)$ matter gauge theories and black holes. arXiv:2208.14031 [hep-th]

54. D. Gaiotto, J.H. Lee, J. Wu, Integrable Kondo problems. JHEP **04**, 268 (2021). https://doi.org/10.1007/JHEP04(2021)268, arXiv:2003.06694 [hep-th]

55. D. Gaiotto, J.H. Lee, B. Vicedo, J. Wu, Kondo line defects and affine Gaudin models. JHEP **01**, 175 (2022). https://doi.org/10.1007/JHEP01(2022)175, arXiv:2010.07325 [hep-th]

56. J. Wu, Anisotropic Kondo line defect and ODE/IM correspondence. arXiv:2106.07792 [hep-th]

57. J. Suzuki, Functional relations in Stokes multipliers and solvable models related to $U_q(A_n^{(1)})$. J. Phys. A **33**, 3507–3522 (2000). https://doi.org/10.1088/0305-4470/33/17/308, arXiv:hep-th/9910215

58. J. Suzuki, Stokes multipliers, spectral determinants and T-Q relations. arXiv:nlin/0009006

59. K. Ito, H. Shu, ODE/IM correspondence and the Argyres-Douglas theory. JHEP **08**, 071 (2017). https://doi.org/10.1007/JHEP08(2017)071, arXiv:1707.03596 [hep-th]

60. K. Ito, T. Kondo, K. Kuroda, H. Shu, WKB periods for higher order ODE and TBA equations. JHEP **10**, 167 (2021). https://doi.org/10.1007/JHEP10(2021)167, arXiv:2104.13680 [hep-th]

61. P. Dorey, C. Dunning, R. Tateo, Differential equations for general SU(n) Bethe ansatz systems. J. Phys. **A33**, 8427–8442 (2000). https://doi.org/10.1088/0305-4470/33/47/308, arXiv:hep-th/0008039 [hep-th]

62. V.V. Bazhanov, A.N. Hibberd, S.M. Khoroshkin, Integrable structure of W(3) conformal field theory, quantum Boussinesq theory and boundary affine Toda theory. Nucl. Phys. B **622**, 475–547 (2002). https://doi.org/10.1016/S0550-3213(01)00595-8, arXiv:hep-th/0105177

Chapter 2
Exact WKB Analysis and TBA Equations

In the previous chapter, we observed a remarkable and mysterious relation between the Schrödinger equation and the quantum integrable model. In this chapter, we will explore a more detailed structure of the relations based on the exact WKB analysis for the Schrödinger equation with a polynomial potential term. The WKB approximation is a semiclassical method of calculating the wave function that is assumed to be an exponential form, where its amplitude and phase change slowly against the coordinate. The wave function is expanded in the Planck constant $\hbar$, which provides an asymptotic series. The Borel resummation, which transforms the asymptotic series into an analytic function, is a useful method to handle the wave function and its analytic continuation exactly. This approach is called the exact WKB method, which is applied to one-dimensional quantum mechanics in [1–7]. The spectral problem of the Schrödinger equation is formulated as the quantization condition for the exact WKB periods, which are obtained from the Borel resummation of the quantum corrections of the wave function.

We are interested in applying the resurgence method to the WKB series in quantum mechanics, which leads to the Borel resummed WKB series of the wave function. Around certain classical trajectories, one can introduce the integration of the (Borel resummed) WKB series, which is called the (Borel resummed) WKB period or quantum period. The different WKB series of the wave function in each sector of the Stokes phenomenon are related to each other via the connection formula [3, 4], which enables us to derive the exact quantization condition of the WKB periods. Solving the exact quantization condition, one can compute the spectrum exactly. The analytic structure of the WKB periods has been well studied since Voros [3]. It is known that the exact WKB period is characterized by its asymptotic series and the discontinuity formula. Determining the exact WKB periods via these two conditions is known as the Voros' Riemann–Hilbert problem. This problem has been well studied in the quartic case in [3], where WKB periods have to be solved by using the traditional ways and then performing the Borel resummation. However, it is more difficult in

K. Ito and H. Shu, *ODE/IM Correspondence and Quantum Periods*, SpringerBriefs in Mathematical Physics 51, https://doi.org/10.1007/978-981-96-0499-9_2

a generic case. Fortunately, in some special cases, e.g. the monic potential, one can apply a more powerful method named the ODE/IM correspondence, which has been reviewed in the previous chapter.

In this chapter, we will see that certain integral equations can reproduce the asymptotic series and the discontinuity formula of the exact WKB periods. These integral equations appear as the TBA equations in the quantum integrable model, which can be derived from the Y-system by generalizing the ODE/IM correspondence in the previous chapter. The main task of this chapter is to apply this method to the Schrödinger equation with arbitrary polynomial potential [8]. We will see that the exact WKB periods are identified with the Y-functions of the TBA equations. The TBA equations should be modified when we vary the moduli parameters of the potential, which is known as the wall-crossing of the TBA equations. In the meantime, the identification between the Y-function and the exact WKB period will also be modified, which is related to the discontinuity formula of the WKB periods. In particular, in the case of the monomial potential, our TBA equations will reproduce those derived from the Y-system in the previous chapter.

In this chapter, we first review the exact WKB analysis of the Schrödinger equation in Sect. 2.1 and present basic facts of the resurgent quantum mechanics in Sect. 2.2. In Sect. 2.3, we present the exact WKB period formulated by its asymptotics and the discontinuity formula, which can be reproduced by using the TBA equations. We then generalize the ODE/IM correspondence to the Schrödinger equation with an arbitrary polynomial potential in Sect. 2.4, whose resulting TBA equations coincide with those of the exact WKB periods. In Sect. 2.5, we introduce the effective central charge of the TBA equations, which provides strong constraints to the classical WKB periods and their quantum corrections. In Sect. 2.6, we then present the wall-crossing of the TBA equations, which enables us to extend the TBA equations to the whole moduli space. At the end of this chapter, we illustrate the general analysis by applying it to the double well oscillator in Sect. 2.7.

2.1 Exact WKB Analysis of the Schrödinger Equation

We present a brief review of the exact WKB analysis of the time-independent Schrödinger equation of a particle with mass $m = 1/2$ in one-dimensional quantum mechanics[1]:

$$\left(-\hbar^2 \frac{d^2}{dx^2} + V(x) - E \right) \psi(x) = 0, \tag{2.1}$$

where $V(x)$ is the potential and E the energy of the system. It is convenient to write this equation in the form by introducing $p(x) = V(x) - E$:

$$\left(-\hbar^2 \frac{d^2}{dx^2} + p(x) \right) \psi(x) = 0, \tag{2.2}$$

---

[1] See also [9] for a review of the exact WKB analysis.

Here we will regard x as the coordinate of the complex plane, and $\hbar$ also as a complex parameter.

Our ansatz of the wave function of Eq. (2.2) is

$$\psi(x) = \exp\left[\frac{1}{\hbar}\int^x P(x')dx'\right], \quad P(x) = \sum_{n=0}^{\infty} \hbar^n p_n(x),$$
(2.3)

where $P(x)$ is a formal power series in $\hbar$. $P(x)$ satisfies the non-linear differential equation called the Riccati equation:

$$P(x)^2 + \hbar\frac{dP(x)}{dx} = p(x).$$
(2.4)

Decomposing the power series $P(x)$ into the even and odd powers of $\hbar$,

$$P(x) = P_{\text{even}}(x) + P_{\text{odd}}(x),$$
(2.5)

the Riccati equation is separated into the $\hbar$-even and odd parts:

$$P_{\text{even}}^2 + P_{\text{odd}}^2 + \hbar\frac{d(P_{\text{odd}})}{dx} = p(x), \quad 2P_{\text{even}}P_{\text{odd}} + \hbar\frac{dP_{\text{even}}}{dx} = 0.$$
(2.6)

P_{odd} is solved as

$$P_{\text{odd}} = -\frac{\hbar}{2}\frac{d}{dx}\log P_{\text{even}}.$$
(2.7)

We thus express the wave function in terms of P_{even} as

$$\psi(x) = \frac{1}{\sqrt{P_{\text{even}}}}\exp\left[\frac{1}{\hbar}\int^x P_{\text{even}}(x')dx'\right].$$
(2.8)

Substituting the formal expansion (2.3) of $P(x)$ into the Riccati equation, one obtains the recursion relation for p_n:

$$\sum_{j=0}^{n} p_j p_{n-j} + \frac{d}{dx}p_{n-1} = p\delta_{n,0},$$
(2.9)

which is written as

$$p_n = -\frac{1}{2p_0}\left(\frac{d}{dx}p_{n-1} + \sum_{j=1}^{n-1} p_j p_{n-j}\right), \quad n = 1, 2, \ldots.$$
(2.10)

The first few examples of p_n are found to be

$$
p_0 = \left(p(x)\right)^{\frac{1}{2}},
$$

$$
p_1 = -\frac{1}{2p_0}\frac{d}{dx}p_0,
$$

$$
p_2 = \frac{1}{4p_0^2}\frac{d^2}{dx^2}p_0 - \frac{3}{8p_0^3}(\frac{d}{dx}p_0)^2, \ldots . \tag{2.11}
$$

From Eq. (2.7), p_n with odd n is the total derivative. The recursion relation determines p_n with $n > 0$ uniquely when we fix the sign of the root in p_0. Since $p_n(x)$ for fixed x is of order $n!$ [10], the WKB expansion of the wave function is an asymptotic series in $\hbar$.

Let us first consider the bound state spectral problem of the Schrödinger equation (2.1), where the wave function decays at infinity. Here the energy E and $V(x)$ are assumed to be real. See Fig. 2.1 for an example of the potential, in which case the bound state exists. We call the zeros of $V(x) - E$ (or $p(x)$) the **turning points**. Depending on the sign of $V(x) - E$, the set of a real coordinate x is divided into some regions. We call the region of x between two real turning points where the energy E is larger (smaller) than the potential $V(E)$ the classically allowed (classically forbidden) region. In the semiclassical limit, the energy is determined by the **Bohr-Sommerfeld quantization condition**, which is given by the integral of $p_0(x) = \sqrt{V(x) - E}$ over a classically allowed region

$$
\frac{2}{i}\int_{\text{classically allowed region}} p_0 dx \sim 2\pi\hbar(n + \frac{1}{2}), \quad n = 0, 1, \ldots . \tag{2.12}
$$

We now consider the case where the potential $V(x)$ is a polynomial function of x. Let $p(x)$ be an $(r + 1)$-th order polynomial in x, parametrized as $p(x) = \sum_{i=0}^{r+1} u_{r+1-i}x^i$. The turning points can be complex in general. Here we assume that there are $(r + 1)$ distinct turning points. To write down the quantization condition with the complex turning points, it is useful to introduce a Riemann surface called the **WKB curve** Σ, which is defined by

$$
y^2 = p(x) = \sum_{i=0}^{r+1} u_{r+1-i}x^i. \tag{2.13}
$$

The Riemann surface (2.13) represents a hyperelliptic curve, which is a two-sheeted branched cover over the Riemann sphere with branched points [11]. Here the branched points are the zeros of $p(x)$, the turning points of the potential. See Fig. 2.1 for an example. The genus g of the WKB curve (2.13) is given by $g = r/2$ for even r or $g = (r - 1)/2$ for odd r. Let $H_1(\Sigma)$ be the homology group of one-cycles on Σ, which has a canonical basis $\{A_I, B_I\}_{I=1,\ldots,g}$ (Fig. 2.2) with intersection numbers: $\langle A_I, B_J \rangle = -\langle B_J, A_I \rangle = \delta_{IJ}$ and $\langle A_I, A_J \rangle = \langle B_I, B_J \rangle = 0$. The right-hand side of

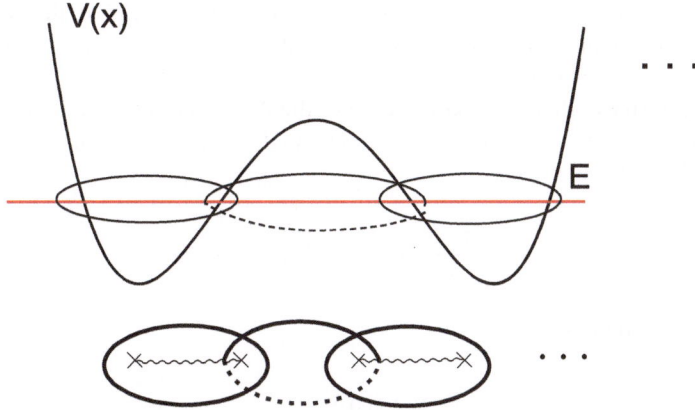

Fig. 2.1 (Top) The potential $p(x)$ and the energy E related to the bound states of the Schrödinger equation. (Bottom) The corresponding WKB curve

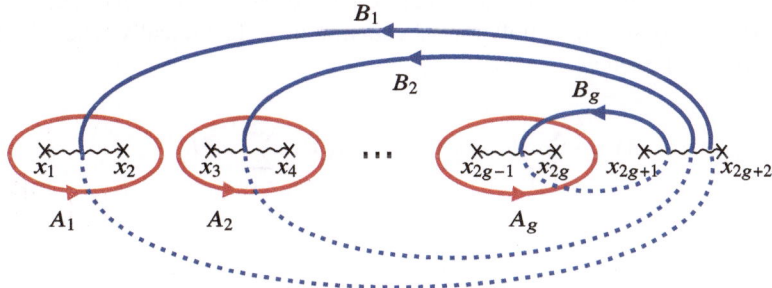

Fig. 2.2 Canonical one-cycles on the hyperelliptic curve

the BS quantization condition (2.12) is written as $\int_\gamma p_0(x)dx$, where γ is the cycle which includes the branch cut corresponding to the allowed region.

We now consider the period integral of $P(x)$ along a one-cycle $\gamma \in H_1(\Sigma)$

$$\Pi_\gamma = \frac{1}{i} \oint_\gamma P(x)dx, \qquad (2.14)$$

which will be denoted by the **WKB period** or the **quantum period**. Substituting the power series (2.3) of $P(x)$, one finds the WKB periods are formal power series of $\hbar^2$

$$\Pi_\gamma(\hbar) = \sum_{n=0}^{\infty} \Pi_\gamma^{(n)} \hbar^{2n} \qquad (2.15)$$

with

$$\Pi_\gamma^{(n)} = \frac{1}{i} \oint_\gamma p_{2n}(x)dx, \quad n = 0, 1, \dots. \qquad (2.16)$$

Since the p_n with odd n can be written by using the total derivative, the coefficients of the odd power of $\hbar$ always vanish. The leading term $\Pi_\gamma^{(0)} = 1/i \int_\gamma p_0(x)dx$ is called the classical WKB period.

To compute the higher-order corrections of the WKB periods, it is useful to express the meromorphic one-form $p_{2n}(x)dx$ on the WKB curve Σ in terms of the basis of meromorphic differentials:

$$\frac{x^{i-1}}{y}dx, \quad i = 1, \ldots, r. \tag{2.17}$$

$p_{2n}(x)$ is expanded as

$$p_{2n}(x)dx = \sum_{j=1}^{r} b_j^{(n)} \frac{x^{i-1}}{y}dx + \partial_x(*)dx, \tag{2.18}$$

where the last term represents the total derivative term. The coefficients $b_j^{(n)}$ can be determined by solving a polynomial equation in x algebraically. Then the WKB periods are expressed as

$$\Pi_\gamma = \sum_{i=1}^{r} b_i(\epsilon)(\Pi_i)_\gamma, \quad b_i(\epsilon) = \sum_{n=0}^{\infty} \epsilon^{2n} b_i^{(2n)}. \tag{2.19}$$

Here

$$(\Pi_i)_\gamma = \int_\gamma \frac{x^{i-1}}{y}dx \tag{2.20}$$

is the period integral of the basis of meromorphic differentials. Properties of asymptotic series are now encoded in the series $b_i(\epsilon)$.

For example, we will present the result for the quartic potential case

$$y^2 = x^4 - u_1 x^2 - u_2 x - u_3. \tag{2.21}$$

The coefficients $b_i^{(n)}$ for $n = 1, 2$ are calculated as [8]

$$b_1^{(1)} = \frac{4(u_1^4 - 9u_1u_2^2 + 16u_1^2u_3 + 48u_3^2)}{3\Delta},$$

$$b_2^{(1)} = 0, \tag{2.22}$$

$$b_3^{(1)} = \frac{3u_1^2u_2^2 - 16u_1^3u_3 + 36u_2^2u_3 - 64u_1u_3^2}{3\Delta},$$

$$b_1^{(2)} = \frac{1}{45\Delta^3} 2\Big(1792u_1^{13} - 39768u_1^{10}u_2^2 + 316302u_1^7u_2^4 - 1022301u_1^4u_2^6 + 1137240u_1u_2^8$$
$$+ 40800u_1^{11}u_3 - 598896u_1^8u_2^2u_3 + 2354760u_1^5u_2^4u_3 - 2599128u_1^2u_2^6u_3 + 390528u_1^9u_3^2$$
$$- 3069696u_1^6u_2^2u_3^2 + 10014624u_1^3u_2^4u_3^2 - 15688080u_2^6u_3^2 + 1555456u_1^7u_3^3$$
$$- 12363264u_1^4u_2^2u_3^3 + 29310336u_1u_2^4u_3^3 + 405504u_1^5u_3^4 - 9934848u_1^2u_2^2u_3^4$$
$$- 12656640u_1^3u_3^5 + 74207232u_2^2u_3^5 - 23101440u_1u_3^6\Big),$$

$$b_2^{(2)} = 0,$$

$$b_3^{(2)} = \frac{1}{\Delta^3}\frac{1}{45}(-1344u_1^{11}u_2^2 + 26676u_1^8u_2^4 - 185139u_1^5u_2^6 + 355752u_1^2u_2^8 + 1792u_1^12u_3$$
$$- 61296u_1^9u_2^2u_3 + 611736u_1^6u_2^4u_3 - 500904u_1^3u_2^6u_3 - 2449440u_2^8u_3 + 38784u_1^10u_3^2$$
$$- 958464u_1^7u_2^2u_3^2 - 48096u_1^4u_2^4u_3^2 + 6792336u_1u_2^6u_3^2 + 708096u_1^8u_3^3 - 1660416u_1^5u_2^2u_3^3$$
$$- 432000u_1^2u_2^4u_3^3 + 5582848u_1^6u_3^4 - 25288704u_1^3u_2^2u_3^4 + 58102272u_2^4u_3^4$$
$$+ 16957440u_1^4u_3^5 - 121614336u_1u_2^2u_3^5 + 10321920u_1^2u_3^6 - 21626880u_3^7).$$

Here Δ is the discriminant of the curve, which is given by

$$\Delta = 4u_1^3u_2^2 - 27u_2^4 - 16u_1^4u_3 + 144u_1u_2^2u_3 - 128u_1^2u_3^2 - 256u_3^3. \tag{2.23}$$

The coefficient $\Pi_\gamma^{(n)}$ for large n diverges factorially:

$$\Pi_\gamma^{(n)} \sim (2n)! \tag{2.24}$$

for fixed values of moduli parameters $\{u_i\}$. The WKB period is thus a divergent asymptotic expansion in $\hbar$ and needs to be resummed to promote the series to a meaningful function, which is the main task of the next section.

2.2 Borel Resummation

In this section, we will employ the Borel resummation technique to treat the divergent series (2.15). We will first introduce the Borel resummation for a general asymptotic series, and then apply it to the WKB period.

Let us consider the formal power series expansion of a certain physical quantity $Z(g)$ depending on a coupling constant g. Let us assume that the weak coupling expansion ($g \sim 0$) of $Z(g)$ takes the form:

$$Z(g) = e^{-A/g} \sum_{n=0}^{\infty} a_n g^{n+\alpha}, \tag{2.25}$$

where $\alpha \neq -1, -2, \ldots$. The factor $e^{-A/g}$ in Eq. (2.25) is known as the non-perturbative part, while the sum represents a perturbative part. When the coefficient

a_n grows as the factorial of n for large n, the series (2.25) becomes an asymptotic series. To study the analytic properties of the series, we first introduce the **Borel transformation** of (2.25) by

$$\mathcal{B}[Z](\xi) = \sum_{n=0}^{\infty} \frac{a_n}{\Gamma(n+\alpha)}(\xi - A)^{n+\alpha-1}. \tag{2.26}$$

Here ξ is a complex variable. Eq. (2.26) provides a convergent series with a finite convergence radius in the complex ξ plane. The function defined by (2.26) gives an analytic function in the whole ξ-plane by analytic continuation. We can recover the asymptotic series (2.25) by using the Laplace transformation of (2.26). To see this, we define the **Borel resummation** by

$$s[Z](g) = \int_0^{\infty} e^{-\xi/g}\mathcal{B}[Z](\xi)d\xi. \tag{2.27}$$

The function $Z(g)$ is said to be **Borel summable** if this integral (2.27) is well-defined. From the definition of the gamma function, we obtain

$$g^{n+\alpha} = \int_0^{\infty} e^{-\xi/g}\frac{\xi^{n+\alpha-1}}{\Gamma(n+\alpha)}d\xi, \tag{2.28}$$

from which we see that the Borel resummation reproduces the formal power series of $Z(g)$ for small g when it is Borel summable.

Since we are working in the complex plane of ξ, it is useful to introduce the **directional Borel resummation** along the direction φ, where the contour goes from 0 to ∞ with the argument fixed by φ,

$$s_\varphi[Z](g) = \int_0^{e^{i\varphi}\infty} e^{-\xi/g}\mathcal{B}[Z](\xi)d\xi. \tag{2.29}$$

$Z(g)$ is called Borel summable along the direction φ if the series $Z(e^{i\varphi}g)$ is Borel summable. If $Z(g)$ is Borel summable, the Borel transform $\mathcal{B}[Z](\xi)$ is an analytic function of ξ with a finite radius of convergence and is extended to the complex ξ-plane by analytic continuation. It can have singularities (poles and/or cuts) in the ξ plane. When the contour of the Laplace transformation (2.27) lies on the singularities, the integral is not well-defined. If the $\mathcal{B}[Z](\xi)$ has a singularity along the φ direction in the Borel plane, we define the lateral Borel resummation by deforming the contour going just above or below the singularities:

$$s_{\varphi^\pm}[Z](g) = \lim_{\delta\to 0_+} s_{\varphi\pm\delta}[Z](g). \tag{2.30}$$

See Fig. 2.3 for the case of $\varphi = 0$.

Fig. 2.3 The singularity of
the Borel transformed series
and the two related contours
of the Laplace
transformation

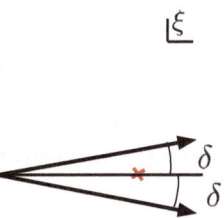

In general, the two limits $s_{\varphi^\pm}$ take different values, which means that s_φ shows a **discontinuity** along the φ-direction, which is defined by

$$\mathrm{disc}_\varphi\big[Z\big](g) = s_{\varphi^+}[Z](g) - s_{\varphi^-}[Z](g). \tag{2.31}$$

In Sect. 2.1, we have introduced the asymptotic series of the wave function and the WKB periods. We now apply the Borel transformation to the WKB period Π_γ associated with the one-cycle γ on the WKB curve by setting $Z = \hbar\Pi_\gamma$, $g = \hbar$ and $\alpha = 1$ in Eq. (2.15)[2]

$$\mathcal{B}[\hbar\Pi_\gamma](\xi) = \sum_{n\geq 0} \frac{1}{(2n)!} \Pi_\gamma^{(n)} \xi^{2n}. \tag{2.32}$$

Here ξ is a complex variable. The Borel resummation of Π_γ is defined by

$$s[\Pi_\gamma](\hbar) = \frac{1}{\hbar} \int_0^\infty e^{-\xi/\hbar} \mathcal{B}[\hbar\Pi_\gamma](\xi) d\xi. \tag{2.33}$$

We also define the Borel resummation along the direction φ in the complex ξ-plane by

$$s_\varphi[\Pi_\gamma](\hbar) = \frac{1}{\hbar} \int_0^{e^{i\varphi}\infty} e^{-\xi/\hbar} \mathcal{B}[\hbar\Pi_\gamma](\xi) d\xi, \tag{2.34}$$

which is equivalent to considering the Laplace transformation defined on the positive real axis but with complex $e^{i\varphi}\hbar$. The discontinuity of the Borel resummation of the Π_γ is defined by

$$\mathrm{disc}_\varphi[\Pi_\gamma](\hbar) = s_{\varphi^+}[\Pi_\gamma](\hbar) - s_{\varphi^-}[\Pi_\gamma](\hbar). \tag{2.35}$$

In the following, we will discuss the singularities of the Borel-transformed WKB periods for generic polynomial potential. We can fix the canonical one-cycles on the WKB curve and study the discontinuity of the Borel resummed WKB periods. There is a formula to determine the discontinuity of the Borel resummed WKB periods in a closed form. Such a formula is called the **discontinuity formula**, presented by Delabare–Pham [7], later reformulated by Iwaki–Nakanishi [12].

---

[2] $\mathcal{B}(\hbar\Pi_\gamma)(\xi)$ is denoted as $\widehat{\Pi}_\gamma(\xi)$ in [8].

2.2.1   Discontinuity Formula of the WKB Period

To derive the discontinuity formula of the WKB periods, one should take advantage of the connection formula of the wave function. Substituting the expansion of $P(x)$ into $\psi(x)$, one can express the wave function (2.8) as[3]

$$\hbar^{1/2}\psi_\pm = \exp\left(\pm\frac{1}{\hbar}\int_{x_0}^x \sqrt{p}\,dx\right)\sum_{n=0}^\infty \psi_\pm^{(n)}(x)\hbar^{n+\frac{1}{2}}. \tag{2.36}$$

Here we specify the initial and end points in the integration. We perform the Borel resummation on the above expansion. We first take the Borel transformation

$$\mathcal{B}[\hbar^{1/2}\psi_\pm](\xi) = \sum_{n=0}^\infty \frac{\psi_\pm^{(n)}}{\Gamma(n+\frac{1}{2})}(\xi \pm s(x))^{n-1/2}, \tag{2.37}$$

where

$$s(x) = \int_{x_0}^x \sqrt{p}\,dx. \tag{2.38}$$

The Borel resummation along the direction φ is

$$\Psi_\pm(\hbar) := s_\varphi[\hbar^{1/2}\psi_\pm](\hbar) = \int_{\mp s(x)}^{\infty e^{i\varphi}} e^{-\xi/\hbar}\mathcal{B}[\psi_\pm]d\xi. \tag{2.39}$$

We then introduce the **Stokes line**. It is the curve starting from a turning point x_0, which satisfies

$$\mathrm{Im}\left(\frac{1}{\hbar}s(x)\right) = \mathrm{Im}\left(\frac{1}{\hbar}\int_{x_0}^x \sqrt{p(x)}\,dx\right) = 0. \tag{2.40}$$

When x moves along the Stokes line, the exponential factor in the wave function $\psi_\pm$ decreases or increases without oscillation. Here we consider the line on the first sheet of the complex plane where the plus sign of the root of $p(x)$ is taken. On the second sheet, where the minus sign is chosen, the behavior of $\psi_\pm$ exchanges. For a simple turning point, which corresponds to a simple zero of $p(x)$, three Stokes lines start from x_0, and each line will end at infinity or another turning point. The graph composed of all the Stokes lines is called the **Stokes graph**.[4] The region with the Stokes line as its boundary is called the **Stokes region**.

In Fig. 2.4, we present two typical examples of the Stokes lines around a simple turning point x_0. Three Stokes lines (solid lines) start from the turning point and end at infinity. If $\mathrm{Re}[\frac{1}{\hbar}s(x)]$ increases (decreases) along the line, ψ_+ (resp. ψ_-) is dominant at infinity, which will be labeled by $+$ (resp. $-$) at the end of the line.

---

[3] Here we have rescaled the wave function by a factor $\hbar^{1/2}$ to follow the convention of [12].

[4] The Stokes graph can be drawn using Mathematica code [13].

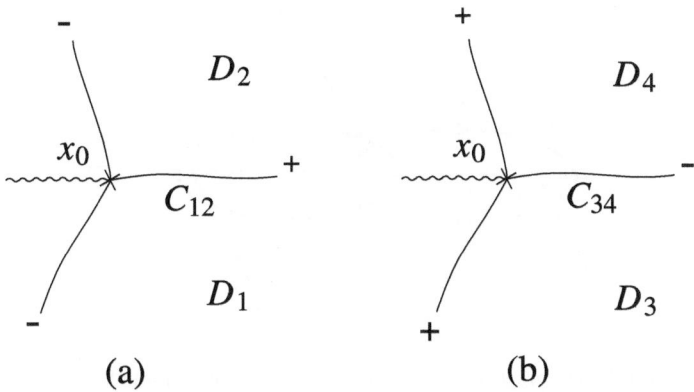

Fig. 2.4 Two typical examples of the Stokes lines around a turning point x_0. Here we have assumed three Stokes lines which end at infinity. The signature $\pm$ at the end of a Stokes line shows that $\psi_\pm$ is dominant on the line

The WKB expansion (2.36) is Borel summable in each Stokes region (see sect. 2 of [12]). Crossing the Stokes line, there is a discontinuity in the wave function, known as the Stokes phenomenon. In the setup of Fig. 2.4, the jumps of the Borel resummed $\Psi_\pm^D$ defined in a Stokes region D are described as follows [3, 4]:

- Crossing the ψ_+-dominant Stokes line C_{12} in the anti-clockwise direction, from the Stokes region D_1 to D_2 in Fig. 2.4:

$$\Psi_+^{D_1} = \Psi_+^{D_2} + i\Psi_-^{D_2}, \quad \Psi_-^{D_1} = \Psi_-^{D_2}. \tag{2.41}$$

- Crossing the ψ_--dominant Stokes line C_{34} in the anti-clockwise direction, from the Stokes region D_3 to D_4 in Fig. 2.4:

$$\Psi_-^{D_3} = \Psi_-^{D_4} + i\Psi_+^{D_4}, \quad \Psi_+^{D_3} = \Psi_+^{D_4}. \tag{2.42}$$

- The jump in the inverse direction is obtained by replacing i with $-i$.

The discontinuity of the WKB wave function leads to that of the Borel resummed WKB periods. Following [12], we consider the Stokes graph in Fig. 2.5 with positive real $\hbar$, as an example. A finite Stokes line that connects two turning points appears. Let γ_0 be a one-cycle encircling the finite Stokes line and γ be a one-cycle that intersects with γ_0. To apply the connection formula of the wave functions, we rotate the phase δ of $\hbar$ sightly such that two different Stokes graphs in Fig. 2.6 appear according to the signs of the imaginary part of $\hbar$. The Stokes regions $D_i, i = 1, 2$, in Fig. 2.5 corresponds to the regions $D_i^\pm$ in Fig. 2.6. We denote $\Psi_{\pm,x_i}^{D_i^\pm}(x)$ by the Borel resummed WKB wave function (2.8) normalized at the turning point x_i in the region $D_i^\pm$:

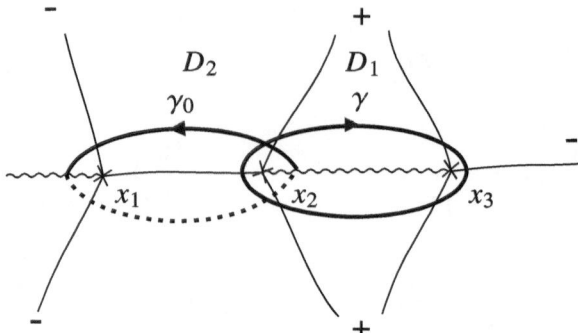

Fig. 2.5 A typical Stokes graph in the derivation of discontinuity formula. The crosses are the turning points. Starting from a turning point x_i, three Stokes lines radiate. The wavy line denotes the branch cut of the WKB curve. The cycle γ_0 encircles the finite Stokes line, while the cycle γ encircles the branch cut

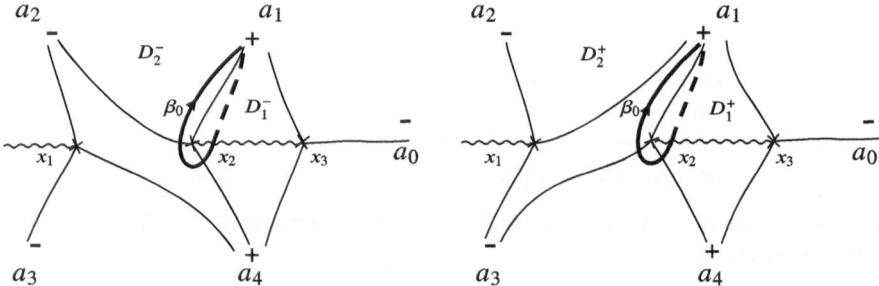

Fig. 2.6 Two Stokes graphs obtained by rotating Fig. 2.5, which lead to two different connection problems from $D_1^\pm$ to $D_2^\pm$. a_i is the point at infinity where the Stokes line ends at and $\psi_\pm$ dominates. β_0 denotes the path around a_1 and turning point x_2

$$\Psi_{\pm,x_i}^{D_i^\pm}(x) := s\left[\frac{\hbar^{1/2}}{\sqrt{P_{\text{even}}(x)}}\exp\left(\pm\frac{1}{\hbar}\int_{x_i}^x P_{\text{even}}(x)dx\right)\right]. \qquad (2.43)$$

When we draw the cycle γ in two graphs of Fig. 2.6, it intersects with the Stokes lines emanating from x_2 differently. To derive the discontinuity formula of the WKB periods, let us consider the connection problems between regions $D_1^\pm$ and $D_2^\pm$. For the left Stokes graph of Fig. 2.6, the connection formula (2.41) from D_1^- to D_2^- leads to

$$\begin{pmatrix} \Psi_{+,x_2}^{D_1^-} \\ \Psi_{-,x_2}^{D_1^-} \end{pmatrix} = \begin{pmatrix} 1 & i \\ 0 & 1 \end{pmatrix}\begin{pmatrix} \Psi_{+,x_2}^{D_2^-} \\ \Psi_{-,x_2}^{D_2^-} \end{pmatrix}. \qquad (2.44)$$

In the connection problem from D_1^+ to D_2^+ for the right Stokes graph of Fig. 2.6, the solution crosses the two Stokes lines, firstly the line emanating from the point x_2 and secondly the one emanating from the point x_1. For the second connection problem, it is convenient to introduce the WKB solution $\Psi_{\pm,x_1}$ normalized at x_1, which is related to $\Psi_{\pm,x_2}$ by

$$\Psi_{\pm,x_2}^{D_2^+}(x) = s\left[e^{\pm\frac{1}{\hbar}\int_{x_2}^{x_1} P_{\text{even}}dx}\right]\Psi_{\pm,x_1}^{D_2^+}(x).$$ (2.45)

Note that the integral is written as

$$\int_{x_2}^{x_1} P_{\text{even}}(x)dx = \frac{1}{2}\int_{\gamma_0} P_{\text{even}}(x)dx.$$ (2.46)

Using the connection formula (2.41), it is found that

$$\begin{pmatrix}\Psi_{+,x_2}^{D_1^+}\\\Psi_{-,x_2}^{D_1^+}\end{pmatrix} = \begin{pmatrix}1 & i\left(1+s\left[\exp(\frac{1}{\hbar}\int_{\gamma_0}P_{\text{even}}(x)dx)\right]\right)\\0 & 1\end{pmatrix}\begin{pmatrix}\Psi_{+,x_2}^{D_2^+}\\\Psi_{-,x_2}^{D_2^+}\end{pmatrix}.$$ (2.47)

We then compare two connection formulas by rewriting them in terms of the WKB solutions $\Psi_{\pm,a_1}(x)$ normalized at the cutoff a_1 in Fig. 2.6, which should be sent to infinity. They are related by

$$\Psi_{\pm,x_i}(x) = \exp\left(\pm\frac{1}{2\hbar}\int_{x_i}^{a_1} P_{\text{even}}(x)dx\right)\Psi_{\pm,a_1}(x), \quad i = 1, 2,$$ (2.48)

where the integral is expressed as that over the cycle β_0 in Fig. 2.6:

$$\int_{x_i}^{a_1} P_{\text{even}}(x)dx = \frac{1}{2}\int_{\beta_0} P_{\text{even}}(x)dx.$$ (2.49)

The connection formulas (2.44) and (2.47) thus can be rewritten as

$$\begin{pmatrix}\Psi_{+,a_1}^{D_1^-}\\\Psi_{-,a_1}^{D_1^-}\end{pmatrix} = \begin{pmatrix}1 & is_{0-}\left[e^{-\frac{1}{\hbar}\int_{\beta_0}P_{\text{even}}(x)dx}\right]\\0 & 1\end{pmatrix}\begin{pmatrix}\Psi_{+,a_1}^{D_2^-}\\\Psi_{-,a_1}^{D_2^-}\end{pmatrix},$$ (2.50)

$$\begin{pmatrix}\Psi_{+,a_1}^{D_1^+}\\\Psi_{-,a_1}^{D_1^+}\end{pmatrix} = \begin{pmatrix}1 & i\left(1+s_{0+}\left[e^{\frac{1}{\hbar}\int_{\gamma_0}P_{\text{even}}(x)dx}\right]\right)s_{0+}\left[e^{-\frac{1}{\hbar}\int_{\beta_0}P_{\text{even}}(x)dx}\right]\\0 & 1\end{pmatrix}\begin{pmatrix}\Psi_{+,a_1}^{D_2^+}\\\Psi_{-,a_1}^{D_2^+}\end{pmatrix}.$$ (2.51)

One can do an analytic continuation from a neighborhood of a_1 to a point in $D_{1,2}^\pm$ along the path that does not intersect with the Stokes line. Then we obtain $\Psi_{\pm,a_1}^{D_{1,2}^+} = \Psi_{\pm,a_1}^{D_{1,2}^-}$. Then Eqs. (2.50) and (2.51) lead to the relation

$$s_{0+}\left[e^{-\frac{1}{\hbar}\int_{\beta_0}P_{\text{even}}(x)dx}\right] = s_{0-}\left[e^{-\frac{1}{\hbar}\int_{\beta_0}P_{\text{even}}(x)dx}\right]\left(1+s\left[e^{\frac{1}{\hbar}\int_{\gamma_0}P_{\text{even}}(x)dx}\right]\right)^{-1}.$$ (2.52)

This formula shows the discontinuity of the integral $\int_{\beta_0} P_{\text{even}}$, where β_0 intersects with γ_1 with the intersection number $\langle \gamma_0, \beta_0 \rangle = -1$. For any path which does not intersect with β_0 the integral does not show discontinuity. The path β_0 can be decomposed as

$$\beta_0 = \beta_{\tilde{a}_1 a_4} + \beta_{a_4, a_2} + \beta_{a_2, a_1}, \tag{2.53}$$

where $\tilde{a}_i$ denotes the point on the second sheet of the WKB curve, which corresponds to p_1. Since only the path β_{a_4, a_2} intersects with the finite Stokes line or the cycle γ_0, we find

$$s_{0^+} \left[e^{-\frac{1}{\hbar} \int_{\beta_{a_4, a_2}} P_{\text{even}} dx} \right] = s_{0^-} \left[e^{-\frac{1}{\hbar} \int_{\beta_{a_4, a_2}} P_{\text{even}} dx} \right] \left(1 + s[e^{\frac{1}{\hbar} \int_{\gamma_0} P_{\text{even}} dx}] \right)^{-1}. \tag{2.54}$$

The one-cycle γ in Fig. 2.5 is also decomposed as $\cdots + \beta_{\cdot a_4} + \beta_{a_4, a_2} + \beta_{a_2, a_1} + \beta_{a_1, \cdot} + \cdots$. Then the discontinuity of the period for γ is written as

$$s_{0^+} \left[e^{-\frac{1}{\hbar} \int_{\gamma} P_{\text{even}}(x) dx} \right] = s_{0^-} \left[e^{-\frac{1}{\hbar} \int_{\gamma} P_{\text{even}}(x) dx} \right] \left(1 + s \left[e^{\frac{1}{\hbar} \int_{\gamma_0} P_{\text{even}}(x) dx} \right] \right)^{\langle \gamma_0, \gamma \rangle}. \tag{2.55}$$

The discontinuity of the WKB periods is written in terms of the **Voros symbol** $\mathcal{V}_\gamma$ $(\gamma \in H_1(\Sigma))$[5]:

$$\mathcal{V}_\gamma = s \left[\exp \left(\frac{1}{\hbar} \int_\gamma P_{\text{even}}(x) dx \right) \right]. \tag{2.56}$$

One of the main applications of the exact WKB method is the **exact quantization condition** of the WKB periods of the Schrödinger equation, which determines the energy spectrum exactly. The exact quantization condition of the Schrödinger equations with double well potential was first conjectured in [14–16] by using the instanton calculus. Given the boundary condition that WKB solutions decrease along two different rays on the complex plane, one can use the connection formula to find the WKB solutions in different regions [6, 17], which finally leads to the exact quantization condition conjectured by using the instanton method. See also [18] for a review.

Note that the Bohr–Sommerfeld quantization condition (2.12) can be expressed by the semiclassical limit of $\mathcal{V}_\gamma = -1$, where γ is the one-cycle associated with the classically allowed region. The exact quantization condition usually involves all the Voros symbols associated with the WKB periods. The condition typically has the form

$$Q(\mathcal{V}_{\gamma_1}, \mathcal{V}_{\gamma_2}, \ldots, \mathcal{V}_{\gamma_r}) = 0. \tag{2.57}$$

Here the Voros symbol has to be replaced by the lateral Borel resummations if the Voros symbol is not Borel summable, where the exact quantization condition may depend on the choice of the lateral Borel resummations.

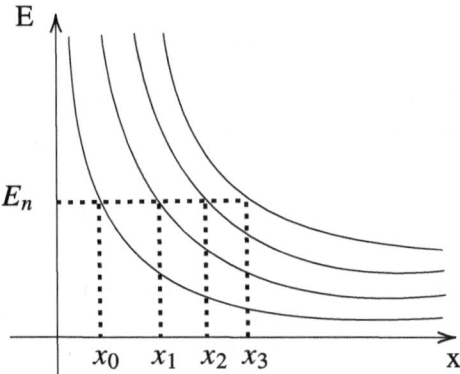

Fig. 2.7 Voros spectrum: eigenvalue in terms of $x = 1/\hbar$

The Voros symbol depends on $\{\hbar, E\}$ and the parameters appearing in $V(x)$. In the standard case, one will fix the Planck constant and the potential $V(x)$, and obtain discrete values of energy E by solving the exact quantization condition. One can also reverse the problem by fixing the potential $V(x)$ and E to derive the discrete values of the inverse Planck constant $x = 1/\hbar$. We will call this viewpoint of the spectrum as **Voros spectrum**, which is a natural form of the spectrum in our study. In Fig. 2.7, we show the discrete values x_n for given energy E_n. In Sect. 2.7, we will combine the TBA equations and the exact quantization condition to solve the Voros spectrum.

2.3 Exact WKB Periods and TBA Equations for Polynomial Potential

In this section, we apply the discontinuity formula for the Schrödinger equation with a polynomial potential and the integral equations satisfied by the Voros symbol.

Let us consider the Schrödinger equation (2.1) with an $(r + 1)$-th order polynomial potential $V(x) = \sum_{a=0}^{r} u_a x^{r+1-a}$, where the polynomial $p(x)$ in (2.2) takes the form $p(x) = V(x) - E = \sum_{a=0}^{r+1} u_a x^{r+1-a}$ with $u_{r+1} = -E$. Here we start with the case where all the turning points are simple and real. We denote the turning points x_i $(i = 1, \ldots, r + 1)$ which are ordered as

$$x_1 < x_2 < \cdots < x_{r+1}. \tag{2.58}$$

On the WKB curve $y^2 = p(x)$, we define the one-cycle γ_i $(i = 1, \ldots, r)$ encircling the turning points x_i and x_{i+1} as in Fig. 2.8. We also choose the potential such that the γ_{2i-1} (γ_{2i}) is associated with the classically allowed (classically forbidden) region, which is possible for $u_0 > 0$. The orientation of γ_i is also indicated in Fig. 2.8 such that their nontrivial intersection number is given by $\langle \gamma_i, \gamma_{i+1} \rangle = (-1)^{i-1}$.

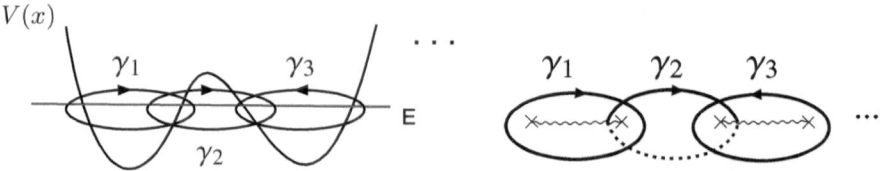

Fig. 2.8 The polynomial potential $V(x)$ where all the turning points are real, and the corresponding WKB curve

On the WKB curve, we introduce the branch cut between x_{2i-1} and x_{2i} ($i = 1, \ldots, [r/2]$). For even r, the cut from x_{r+1} to $+\infty$ is also defined. We have chosen the branch cuts and the orientations of the cycles such that[5]

$$m_{2i-1} = \Pi^{(0)}_{\gamma_{2i-1}} = \frac{2}{i} \int_{x_{2i-1}}^{x_{2i}} \sqrt{p(x)}dx, \quad m_{2i} = i\Pi^{(0)}_{\gamma_{2i}} = 2 \int_{x_{2i}}^{x_{2i+1}} \sqrt{p(x)}dx \quad (2.59)$$

are real and positive. Let us take γ_i as a cycle corresponding to γ_0 in Fig. 2.5. γ_i intersects with γ_{i-1} and γ_{i+1}. For even i, the finite Stokes line starting from x_i to x_{i+1} appears for $\varphi = \arg(\hbar) = 0, \pi$. For odd i, the finite Stokes line appears for $\varphi = \frac{\pi}{2}, \frac{3\pi}{2}$. According to the discontinuity formula (2.55), the WKB period $\Pi_{\gamma_{2i}}$ is Borel summable along the direction $\varphi = 0$. The WKB period $\Pi_{\gamma_{2i-1}}$ is not Borel summable along the direction $\varphi = 0, \pi$, whose discontinuity can be expressed by using the WKB period $\Pi_{\gamma_{2i-2}}$ and $\Pi_{\gamma_{2i}}$:

$$\mathrm{disc}_{0,\pi}[\Pi_{\gamma_{2i-1}}](\hbar) = i\hbar \log\left(1 + \exp\left(-\frac{i}{\hbar}\Pi_{\gamma_{2i-2}}\right)\right) + i\hbar \log\left(1 + \exp\left(-\frac{i}{\hbar}\Pi_{\gamma_{2i}}\right)\right), \quad (2.60)$$

where we have taken the logarithm of Eq. (2.55). We have a similar discontinuity along the direction $\varphi = \pm\pi/2$ for $\Pi_{\gamma_{2i}}$:

$$\mathrm{disc}_{\pm\pi/2}[\Pi_{\gamma_{2i}}](\hbar) = -i\hbar \log\left(1 + \exp\left(-\frac{i}{\hbar}\Pi_{\gamma_{2i-1}}\right)\right) - i\hbar \log\left(1 + \exp\left(-\frac{i}{\hbar}\Pi_{\gamma_{2i+1}}\right)\right). \quad (2.61)$$

The singularities of the Borel resummed WKB periods occur only on the rays with the above angles and their discontinuities are characterized by Eqs. (2.60) and (2.61). The problem of recovering the function from the data around the singularities is called the Riemann–Hilbert problem. The Riemann–Hilbert problem for the WKB periods has been formulated in [3], where the problem is solved for the quartic potential case. For general polynomial potential, however, it is very complicated to study the Borel resummation of the WKB periods and their analytic structure. In [8], the discontinuity formula is written in terms of the TBA equation that appears in Chap. 1, whose solution provides the Borel resummation of the WKB periods with expected analytic properties.

To obtain the TBA equation satisfied by the WKB periods, we introduce the rapidity $\theta = -\log \hbar$ and define the pseudo energy by

---

[5] Note that γ_1 and γ_3 in Fig. 2.8 have different signs for each Stokes line.

$$\epsilon_{2i-1}\left(\theta + \frac{\pi i}{2} \mp i0_+\right) := \frac{i}{\hbar}s_\pm[\Pi_{\gamma_{2i-1}}](\hbar), \quad \epsilon_{2i}(\theta) := \frac{i}{\hbar}s[\Pi_{\gamma_{2i}}](\hbar). \tag{2.62}$$

Here 0_+ represents the infinitesimally positive small number. Note that the argument of the first equation is shifted by $i\frac{\pi}{2}$, which is useful to unify the discontinuity formulas (2.60) and (2.61) as

$$\mathrm{disc}_{\pi/2}[\epsilon_a](\theta) = L_{a-1}(\theta) + L_{a+1}(\theta), \quad a = 1, \dots, r, \tag{2.63}$$

where $L_0 = L_{r+1} = 0$ and $L_a(\theta)$ is defined by

$$L_a(\theta) = \log\left(1 + e^{-\epsilon_a(\theta)}\right), \quad a = 1, \dots, r. \tag{2.64}$$

The asymptotic behavior of the pseudo energy at $\theta \to \infty$ corresponding to the semiclassical limit $\hbar \to 0$ is given by

$$\epsilon_a(\theta) \sim m_a e^\theta + O(e^{-\theta}), \quad \theta \to \infty, \tag{2.65}$$

where m_a is defined in (2.59). We should thus find the analytic functions $\epsilon_a(\theta)$, which satisfy the discontinuity (2.63) and the semiclassical limit (2.65). The solution to this problem is found to satisfy the TBA equation [8]:

$$\epsilon_a(\theta) = m_a e^\theta - \int_{-\infty}^\infty \frac{d\theta'}{2\pi} \frac{L_{a-1}(\theta') + L_{a+1}(\theta')}{\cosh(\theta - \theta')}, \tag{2.66}$$

which will also be shown by using the ODE/IM correspondence in the next section. Here we will show that the discontinuity formula follows from the TBA Eq. (2.63). Shifting variable θ by $\pi i/2 \pm i0_+$, Eq. (2.63) becomes

$$\epsilon_a\left(\theta + \frac{\pi i}{2} \pm i0_+\right) = im_a e^{\theta \pm i0_+} - \int_{-\infty}^\infty \frac{d\theta'}{2\pi} \frac{L_{a-1}(\theta') + L_{a+1}(\theta')}{i\sinh(\theta - \theta' \pm i0_+)}, \tag{2.67}$$

which leads to the discontinuity formula (2.63):

$$\epsilon_s\left(\theta + \frac{\pi i}{2} + i0_+\right) - \epsilon_s\left(\theta + \frac{\pi i}{2} - i0_+\right) = L_{s-1}(\theta) + L_{s+1}(\theta). \tag{2.68}$$

Here we have used the formula

$$\frac{1}{2\pi \cosh(\theta - \theta' + \frac{i\pi}{2} - i0_+)} - \frac{1}{2\pi \cosh(\theta - \theta' + \frac{i\pi}{2} + i0_+)} = \delta(\theta - \theta'). \tag{2.69}$$

Note that the TBA equation of the quantum integrable model can be rewritten in the form of the Y-system under assumptions of certain analytic properties. In the previous chapter, we derived the Y-system from the Schrödinger equation with a monomial potential by using the ODE/IM correspondence. In the next section, we will derive the TBA Eq. (2.66) by generalizing the ODE/IM correspondence to the case of a polynomial potential.

2.4 ODE/IM Correspondence for Polynomial Potential

We now study the ODE/IM correspondence for the ODE (1.1) with a polynomial potential [8]. Let us consider the ODE

$$
\left(-\frac{d^2}{dz^2} + z^{r+1} + \sum_{a=1}^{r} b_a z^{r-a} \right) \psi(z, b_a) = 0
\tag{2.70}
$$

defined in the complex z-plane, which can be obtained by letting $u_0 = 1$ in the Schrödinger equation (2.2) and rescaling x such that the Planck constant is scaled to 1. b_a $(a = 1, \ldots, r)$ are complex coefficients. r is an integer larger than one. Eq. (2.70) is invariant under the rotation of the variables [19]

$$
(z, b_a) \rightarrow (\omega z, \omega^{a+1} b_a), \quad \omega = e^{\frac{2\pi i}{r+3}}.
\tag{2.71}
$$

This rotation generates new solutions from a given solution of the ODE. Let $y(z, b_a)$ be the solution that decays at infinity along the positive real axis. The WKB analysis shows that its asymptotic is given by

$$
y(z, b_a) \sim \frac{1}{\sqrt{2i}} z^{n_r} \exp\left(-\frac{2}{r+3} z^{\frac{r+3}{2}} \right),
\tag{2.72}
$$

where n_r are defined by

$$
n_r =
\begin{cases}
-\frac{r+1}{4} & r+1 : \text{odd} \\
-\frac{r+1}{4} - B_{\frac{r+3}{2}} & r+1 : \text{even}
\end{cases},
\tag{2.73}
$$

where $B_m(b_a)$ $(m = 1, 2, \ldots)$ are defined by

$$
\left(1 + \sum_{a=1}^{r} b_a z^{-(a+1)} \right)^{1/2} =: 1 + \sum_{m=1}^{\infty} B_m z^{-m}.
\tag{2.74}
$$

Note that the coefficient $B_{\frac{r+3}{2}}(b_a)$ satisfies

$$
B_{\frac{r+3}{2}}(\omega^{-k(a+1)} b_a) = (-1)^k B_{\frac{r+3}{2}}(b_a).
\tag{2.75}
$$

The solution (2.72) is subdominant in the sector $\mathcal{S}_0 : |\arg(z)| < \frac{\pi}{r+3}$. Applying the rotation (2.71) k times where k is an integer, we obtain the subdominant solution $y_k(z, b_a)$ in sector $\mathcal{S}_k : |\arg(z) - \frac{2k\pi}{r+3}| < \frac{\pi}{r+3}$:

$$y_k(z, b_a) = \omega^{\frac{k}{2}} y(\omega^{-k} z, \omega^{-(a+1)k} b_a). \tag{2.76}$$

In our normalization of the solution (2.72), the Wronskian (1.9) of y_k and y_{k+1} is

$$W[y_k, y_{k+1}](b_a) = \begin{cases} 1 & r+1 : \text{odd} \\ \omega^{(-1)^k B_{\frac{r+3}{2}}} & r+1 : \text{even} \end{cases}. \tag{2.77}$$

Writing the phase rotation of a function $f(z, b_a)$ by

$$f^{[j]}(z, b_a) = f(\omega^{-j/2} z, \omega^{-j(a+1)/2} b_a), \quad j \in \mathbb{Z}, \tag{2.78}$$

the rotation of the Wronskian of y_k's satisfies the relation

$$W_{k_1+1, k_2+1}(b_a) = W^{[2]}_{k_1, k_2}(b_a), \tag{2.79}$$

where we defined $W_{k_1, k_2} := W[y_{k_1}, y_{k_2}]$.

We now define the T-functions (see Eq. (1.69)) by

$$T_a(b_a) = W^{[-(a+1)]}_{0, a+1}, \quad a \in \mathbb{Z}. \tag{2.80}$$

The T-system for $T_a(b_a)$ takes the same form for even r, but for odd r, it includes complicated factors due to the normalization factor in (2.77). We then introduce the Y-function by taking the ratios of the T-functions

$$\mathcal{Y}_{2j}(b_a) = \frac{W_{-j,j}(b_a) W_{-j-1,j+1}(b_a)}{W_{-j-1,-j}(b_a) W_{j,j+1}(b_a)},$$
$$\mathcal{Y}_{2j+1}(b_a) = \left[\frac{W_{-j-1,j}(b_a) W_{-j-2,j+1}(b_a)}{W_{-j-2,-j-1}(b_a) W_{j,j+1}(b_a)} \right]^{[+1]}, \quad j \in \mathbb{Z}_{\geq 0}. \tag{2.81}$$

The normalization factor (2.77) cancels out in this definition. Using the the Plücker relation (1.22), we find the Y-system:

$$\mathcal{Y}^{[+1]}_s(b_a) \mathcal{Y}^{[-1]}_s(b_a) = \left(1 + \mathcal{Y}_{s-1}(b_a)\right) \left(1 + \mathcal{Y}_{s+1}(b_a)\right), \quad s \in \mathbb{Z}. \tag{2.82}$$

By the definition of the $\mathcal{Y}_s$-function, $\mathcal{Y}_0(b_a) = 0$. Note that $y_{j+r+3}(z)$ and $y_j(ze^{2\pi i})$ are the subdominant solutions in the same sector, we thus find $y_{j+r+3}(z) \propto y_j(ze^{2\pi i})$. Since the monodromy around origin is trivial, $y_{j+r+3}(z) \propto y_j(ze^{2\pi i}) = y_j(z)$, which leads to $\mathcal{Y}_{r+1}(b_a) = 0$. Therefore, it is natural to truncate the system (2.82) at r. We thus obtain an A_r-type Y-system. In the special case $b_1 = b_2 = \cdots = b_{r-1}$, the generalized ODE (2.70) reduces to the ODE (1.1) in the previous chapter. The Y-system (2.82) also reproduces the relation (1.27) in this special limit.

However, the Y-system (2.82) has multiple spectral parameters b_a, which makes it difficult to convert the Y-system to the TBA equations. To avoid this difficulty, we recover the Planck constant denoted by ζ by rescaling

$$x = \zeta^{\frac{2}{r+3}} z, \quad u_a = -\zeta^{\frac{2(a+1)}{r+3}} b_a, \quad a = 1, 2, \ldots, r. \tag{2.83}$$

In the new variables, the ODE (2.70) becomes Eq. (2.2) but with replacing $\hbar$ by ζ. Thanks to the rescaling, the rotation of multiple parameters (2.71) can be rewritten by only rotating ζ, which is now promoted to a single spectral parameter. The subdominant solution $\hat{y}(x, u, \zeta_a)$ of Eq. (2.2) can be obtaining by rescaling $y(z, b_a)$ by

$$\hat{y}(x, u_a, \zeta) := y(z, b_a) = y(\zeta^{-\frac{2}{r+3}} x, -\zeta^{-\frac{2(a+1)}{r+3}} u_a), \tag{2.84}$$

which also generates the new solutions by the rotation of ζ:

$$\hat{y}_k(x, u_a, \zeta) = \omega^{\frac{k}{2}} \hat{y}(x, u_a, e^{i\pi k} \zeta), \quad k \in \mathbb{Z}. \tag{2.85}$$

$\hat{y}_k(x, u_a, \zeta)$ corresponds to $y_k(z, b_a)$ in Eq. (2.76). We introduce the rescaled Wronskian $\hat{W}_{k_1, k_2}$ by

$$\hat{W}_{k_1, k_2}(\zeta, u_a) = \zeta^{\frac{2}{r+3}} W[\hat{y}_{k_1}, \hat{y}_{k_2}] = W_{k_1, k_2}(b_a). \tag{2.86}$$

We then rewrite the Y-function $\mathcal{Y}_s(b_a)$ as $Y_s(\zeta, u_a)$ in terms of new variables (ζ, u_a)[6]

$$Y_{2j}(\zeta, u_a) = \frac{\hat{W}_{-j,j} \hat{W}_{-j-1,j+1}}{\hat{W}_{-j-1,-j} \hat{W}_{j,j+1}}(\zeta, u_a),$$

$$Y_{2j+1}(e^{-\frac{\pi i}{2}} \zeta, u_a) = \frac{\hat{W}_{-j-1,j} \hat{W}_{-j-2,j+1}}{\hat{W}_{-j-2,-j-1} \hat{W}_{j,j+1}}(\zeta, u_a),$$
$$\tag{2.87}$$

which satisfy the (A_1, A_r)-type Y-system:

$$Y_s(i\zeta, u_a) Y_s(-i\zeta, u_a) = \left(1 + Y_{s+1}(\zeta, u_a)\right)\left(1 + Y_{s-1}(\zeta, u_a)\right), \quad s = 0, \ldots, r, \tag{2.88}$$

where $Y_0(\zeta, u_a) = 0 = Y_{r+1}(\zeta, u_a)$. The new Y-system (2.88) shows the same relation as the original one (2.82) but different dependence on the spectral parameters.

We then convert the Y-system (2.88) into a set of TBA equations. We first study the asymptotic behavior of the Y-functions (2.87) at small ζ, where one can use the WKB analysis in ζ. At small ζ, the WKB expansion of $\hat{y}_k(x, u_a, \zeta)$ is

---

[6] The same Y-system also appears in the study of minimal surface in the scattering amplitude/Wilson loop duality [20, 21].

$$\hat{y}_k(x, u_a, \zeta) = (-1)^{\frac{k}{2}} c(\zeta) \exp\left(\frac{\delta_k}{\zeta} \int_{a_k}^{x} P(x')dx'\right), \tag{2.89}$$

where $P(x)$ is given by Eq. (2.3). $c(\zeta) = \frac{1}{\sqrt{2i}} \zeta^{\frac{(r+1)}{2(r+3)}}$ and a_k is the initial point of the integration. $\delta_k = \pm(-1)^k$ means the sign factor, where $\pm$ depends on the sheet of the WKB curve on which x lives. We should choose the minus sign in $\pm$ to match with the large x (large z) asymptotic behaviors of $\hat{y}_k$ that decays at infinity in the sector S_k. When the integration contour starts from a_k and ends with x through the branch cut, one could choose the plus sign in $\pm$.

Keeping these in mind, we evaluate the Wronskian $\hat{W}_{k,j}$ for the solutions $\hat{y}_k$ and $\hat{y}_j$. Substituting the WKB solution (2.89) into the Wronskian (2.86), we get

$$\hat{W}_{k,j} = i(-1)^{\frac{k+j}{2}} \delta_k \exp\left(\frac{\delta_k}{\zeta} \int_{a_k}^{a_j} P_{\text{even}}(x')dx' + \frac{1}{2}[\log P_{\text{even}}(x_j) + \log P_{\text{even}}(x_k)]\right), \tag{2.90}$$

where we have used the decomposition of $P(x)$ in Eq. (2.5). Here we have chosen the sheets such that the explicit x dependence does not appear in the expression. This can be done when the condition $\delta_k = -\delta_j$ holds.

A Y-function is composed of four Wronskians, which provide four integration contours. Combining these four contours, one obtains a closed cycle associated with the Y-function. Let us illustrate this procedure by applying it to the Y-functions $Y_1(\zeta)$ and $Y_3(\zeta)$ for the potential with $r = 3$, where the Y-functions are given by

$$Y_1(e^{-\frac{\pi i}{2}}\zeta) = -\frac{W_{-1,0}W_{1,-2}}{W_{-1,-2}W_{1,0}}, \quad Y_3(e^{-\frac{\pi i}{2}}\zeta) = -\frac{W_{1,-2}W_{-3,2}}{W_{-3,-2}W_{1,2}}. \tag{2.91}$$

The Wronskian $W_{j,k}$ on the numerator (denominator) is represented by an oriented line from a_j to a_k (from a_k to a_j) in Fig. 2.9. Y-function $Y_1(e^{-\frac{\pi i}{2}}\zeta)$ and $Y_3(e^{-\frac{\pi i}{2}}\zeta)$ are finally associated with the cycles encircling the two branch cuts.

Repeating this procedure for other Y-functions (2.87), we find in the limit $\zeta \to 0$, the Y-functions behave as

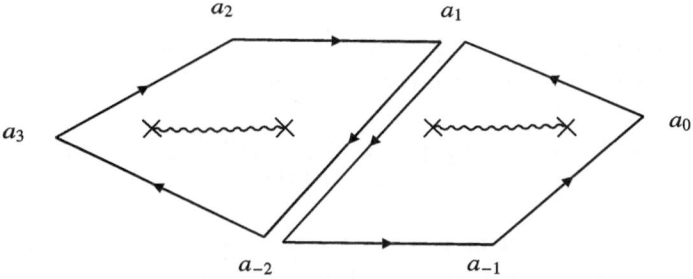

Fig. 2.9 The one-cycles associated with the Y-functions $Y_1(\zeta)$ (right) and $Y_3(\zeta)$ (left) for the potential with $r = 3$

$$\log Y_{2j+1}(\zeta, u_a) \sim \frac{1}{i\zeta} \oint_{\gamma_{r-2j}} p_0(x)dx =: -\frac{m_{r-2j}}{\zeta},$$

$$\log Y_{2j}(\zeta, u_a) \sim \frac{1}{\zeta} \oint_{\gamma_{r+1-2j}} p_0(x)dx =: -\frac{m_{r+1-2j}}{\zeta}, \quad \zeta \to 0 \tag{2.92}$$

in the region $|\arg(\zeta)| < \pi$. Here the cycle γ_i is given as in Fig. 2.8 for the turning points (2.58). m_s are always positive real value in the case of Fig. 2.8, which depends on the moduli space parameter $(u_1, \ldots, u_r)$. There are thus r independent m_s in the present case. However, note that the Y-function Y_k corresponds to the one-cycle γ_{r+1-k}. To avoid this confusion, we will relabel the Y-function by $Y_k \to Y_{r+1-k}$ from now, which leaves the form of the Y-system invariant.

2.4.1   TBA Equations

As we saw in Chap. 1, the TBA equations can be obtained from the Y-system with certain asymptotic conditions for the Y-functions. We now derive the TBA equations from the Y-system (2.88) [20]. For $\zeta \neq 0$, the logarithms of the Y-functions should be analytic functions of ζ in $|\arg(\zeta)| \leq \frac{\pi}{2}$, but become singular at $\zeta = 0$. We introduce $\ell_s(\zeta)$, which is analytic in $|\arg(\zeta)| \leq \frac{\pi}{2}$:

$$\ell_s(\zeta) := \log Y_s(\zeta) + \frac{m_s}{\zeta}. \tag{2.93}$$

Taking the logarithm of the Y-system (2.88), we obtain

$$\ell_s(e^{\frac{\pi i}{2}}\zeta) + \ell_s(e^{-\frac{\pi i}{2}}\zeta) = \log\left((1 + Y_{s+1}(\zeta))(1 + Y_{s-1}(\zeta))\right). \tag{2.94}$$

We then set $\zeta = e^{-\theta}$ and introduce the kernel function

$$K(\theta) = \frac{1}{2\pi \cosh \theta}. \tag{2.95}$$

We take the convolution of Eq. (2.94) with the kernel K. Shifting the contour of the convolutions, the left-hand side of (2.94) becomes

$$K * \left[\ell_s(e^{\frac{\pi i}{2}}\zeta) + \ell_s(e^{-\frac{\pi i}{2}}\zeta)\right] = \oint_{\gamma_*} \frac{d\theta'}{2\pi i} \frac{\ell_s(\theta')}{\sinh(\theta - \theta')} = \ell_s(\theta), \tag{2.96}$$

where the integration contour γ_* is $-\infty + \frac{\pi i}{2} \to \infty + \frac{\pi i}{2} \to \infty - \frac{\pi i}{2} \to -\infty - \frac{\pi i}{2} \to -\infty + \frac{\pi i}{2}$ (Fig. 2.10). Here we have used the analytic property of ℓ_s in $|\arg(\zeta)| \leq \frac{\pi}{2}$, which leads to $\frac{\ell_s(\theta')}{\sinh(\theta-\theta')} \to 0$ at infinity.

Finally, we obtain the TBA equations

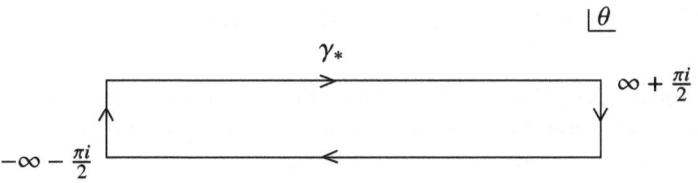

Fig. 2.10 Contour γ_* on the θ plane

$$\log Y_s(\theta) = -m_s e^\theta + K * \log\Big((1 + Y_{s+1})(1 + Y_{s-1})\Big). \qquad (2.97)$$

These TBA equations hold for $\text{Im}(\theta) \leq \frac{\pi}{2}$. The first term in the right-hand side of Eq. (2.97) is called the source term and represents the asymptotics (2.92) of $\log Y_s(\theta)$ in the $\zeta \to 0$ limit. These TBA equations thus become of the form Eq. (2.66) in Sect. 2.3, if we identify

$$\log Y_s(\theta) = -\epsilon_s(\theta). \qquad (2.98)$$

In fact, in terms of ϵ_s and L_s in Eq. (2.64), we can rewrite the TBA equations as

$$\epsilon_s(\theta) = m_s e^\theta - K * \Big(L_{s-1} + L_{s+1}\Big). \qquad (2.99)$$

Note that we have assumed that all the turning points are real so far, which will be extended to the whole moduli space of the potential in Sect. 2.6.

The solution of the TBA equations gives the same singularities and the same asymptotic behavior as the exact WKB periods. We can test the solutions of the TBA equations by comparing the WKB periods with the logarithm of the Y-functions, where we identify $\zeta = \hbar$. At large θ/small $\hbar$, we expand the right-hand side of the TBA equations (2.99) in $e^{-\theta}$, which gives the series

$$\epsilon_s(\theta) \sim m_s e^\theta + \sum_{n=1}^{\infty} m_s^{(n)} e^{(1-2n)\theta}, \qquad (2.100)$$

where the coefficient $m_s^{(n)}$ is

$$m_s^{(n)} = \frac{(-1)^n}{\pi} \int_{-\infty}^{\infty} d\theta \, e^{(2n-1)\theta} \Big(L_{s-1}(\theta) + L_{s+1}(\theta)\Big). \qquad (2.101)$$

On the other hand, the small $\hbar$ expansion (2.15) in Sect. 2.1 can be computed by using the Picard–Fuchs operator. According to the definitions (2.62), we need to test the following identifications:

$$m_{2i-1}^{(n)} = (-1)^n \Pi_{\gamma_{2i-1}}^{(n)}, \quad m_{2i}^{(n)} = i \Pi_{\gamma_{2i}}^{(n)}. \qquad (2.102)$$

In Sect. 2.7, we will test these identifications through the examples of the quartic oscillator. See also [8, 22] for more evidence.

2.5   Effective Central Charge and PNP Relation

The TBA equations (2.99) appear as the kink limit of the TBA equations of a massive integrable model. From the kink limit of the free energy of the massive integrable model [23], one has the **effective central charge**

$$c_{\text{eff}} = \frac{6}{\pi^2} \sum_{s=1}^{r} m_s \int_{-\infty}^{\infty} d\theta \, e^{\theta} L_s(\theta). \tag{2.103}$$

This can be evaluated analytically by using the Rogers dilogarithm identity [24] in a standard way [23, 25]:

$$c_{\text{eff}} = \sum_{s=1}^{r} \frac{6}{\pi^2} \mathcal{L}\left(\frac{1}{1 + e^{\epsilon_s(-\infty)}}\right), \tag{2.104}$$

where $\mathcal{L}(t)$ is Rogers' dilogarithm function

$$\mathcal{L}(t) = -\frac{1}{2} \int_0^t \left(\frac{\log(1 - t')}{t'} + \frac{\log t'}{1 - t'}\right). \tag{2.105}$$

The value of $\epsilon_s(\theta)$ at the limit $\theta \to -\infty$ is given by

$$e^{-\epsilon_s(-\infty)} = \frac{\sin\left(\frac{\pi s}{r+3}\right) \sin\left(\frac{\pi(s+2)}{r+3}\right)}{\sin^2\left(\frac{\pi}{r+3}\right)}. \tag{2.106}$$

One finds the effective central charge

$$c_{\text{eff}} = \frac{r(r+1)}{r+3}. \tag{2.107}$$

The constant value of effective central charge leads to a strong constraint on the classical WKB periods and their quantum corrections. From Eq. (2.101), $m_s^{(1)}$ is written in terms of the integration of $L_{s-1} + L_{s+1}$

$$m_s^{(1)} = -\frac{1}{\pi} \int_{-\infty}^{\infty} d\theta \, e^{\theta} \left(L_{s-1}(\theta) + L_{s+1}(\theta)\right). \tag{2.108}$$

In the case of even r, one can always express the effective central charge c_{eff} as the combinations of m_s and their next order correction. However, it is complicated to present the general expression. Let us consider the case $r = 2$ and $r = 4$ for instance

$$c_{\text{eff}}(r=2) = -\frac{6}{\pi}(m_1 m_2^{(1)} + m_2 m_1^{(1)}),$$

$$c_{\text{eff}}(r=4) = -\frac{6}{\pi}\left(m_1(m_2^{(1)} - m_4^{(1)}) + m_2 m_1^{(1)} + m_3 m_4^{(1)} + m_4(m_3^{(1)} - m_1^{(1)})\right),$$

$$(2.109)$$

which is known as the perturbative and non-perturbative (PNP) relation of the WKB periods [26–30].

Using the identification between masses and the WKB periods (2.59) and (2.102), the constant value of effective central charge leads to the relation between the classical WKB periods and their quantum correction

$$r = 2\text{case}: \quad \frac{i\pi}{5} = \Pi_{\gamma_1}^{(0)}\Pi_{\gamma_2}^{(1)} - \Pi_{\gamma_1}^{(0)}\Pi_{\gamma_1}^{(1)},$$

$$r = 4\text{case}: \quad \frac{10}{21}i\pi = \Pi_{\gamma_1}^{(0)}(\Pi_{\gamma_2}^{(1)} - \Pi_{\gamma_4}^{(1)}) - \Pi_{\gamma_2}^{(0)}\Pi_{\gamma_1}^{(1)} + \Pi_{\gamma_3}^{(0)}\Pi_{\gamma_4}^{(1)} + \Pi_{\gamma_4}^{(0)}(\Pi_{\gamma_1}^{(1)} - \Pi_{\gamma_3}^{(1)}).$$

$$(2.110)$$

In the case of odd r, we do not find this type of constraint in the generic case. We will show a special case of the double well potential in Sect. 2.7.1. These constraints between the classical WKB period and its quantum correction are also known as the quantum version of the Matone relations [31–33].

2.6 Wall-Crossing of TBA Equations for General Polynomial Potential

So far, we have only considered the specific cases, where all the turning points are real and different. Varying the parameters u_a of the polynomial $p(x)$ in Eq. (2.2), some turning points will take a complex value. In this case, the masses m_s in Eqs. (2.59) can be complex. Suppose the phase of the complex mass m_s is ϕ_s. We then shift the variable θ of the TBA Eq. (2.97) such that the source term $m_s e^\theta = |m_s| e^{\theta + i\phi_s}$ is real, i.e. shift $\theta \to \theta - i\phi_s$. Then we obtain the TBA equations for complex masses

$$\log Y_s(\theta - i\phi_s) = -|m_s|e^\theta + \int_{-\infty}^{\infty} d\theta' \frac{\log[1 + Y_{s+1}(\theta' - i\phi_{s+1})]}{2\pi \cosh(\theta - \theta' - i\phi_s + i\phi_{s+1})}$$
$$+ \int_{-\infty}^{\infty} d\theta' \frac{\log[1 + Y_{s-1}(\theta' - i\phi_{s-1})]}{2\pi \cosh(\theta - \theta' - i\phi_s + i\phi_{s-1})}.$$

$$(2.111)$$

We introduce the shifted pseudo-energy function

$$\tilde{\epsilon}_s(\theta) = \epsilon_s(\theta - i\phi_s) = -\log Y_s(\theta - i\phi_s), \quad \tilde{L}_s(\theta) = L_s(\theta - i\phi_s). \quad (2.112)$$

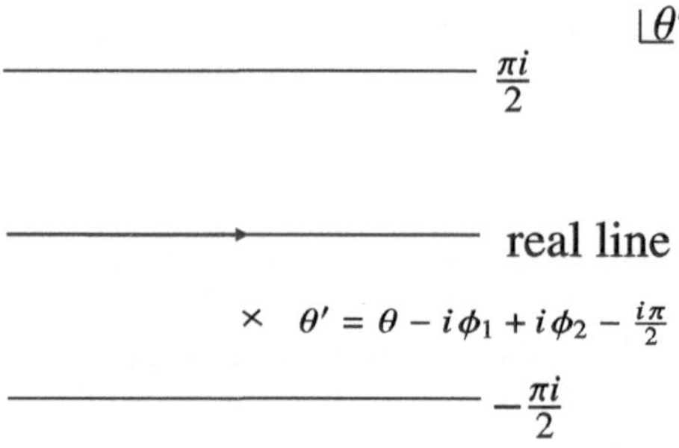

Fig. 2.11 The integration contour (blue line) of the convolution and the location of the pole (crossed point) $\theta' = \theta - i\phi_1 + i\phi_2 - \frac{i\pi}{2}$

The TBA Eq. (2.111) become

$$\tilde{\epsilon}_s = |m_s|e^\theta - K_{s,s-1} * \tilde{L}_{s-1} - K_{s,s+1} * \tilde{L}_{s+1}, \quad s = 1, \ldots, r, \tag{2.113}$$

where $K_{s',s}(\theta)$ the shifted kernel defined by

$$K_{s',s}(\theta) = \frac{1}{2\pi \cosh(\theta - i\phi_{s'} + i\phi_s)}. \tag{2.114}$$

The effective central charge (2.103) becomes

$$c_{\text{eff}} = \frac{6}{\pi^2} \sum_{s=1}^{r} |m_a| \int_{-\infty}^{\infty} d\theta e^\theta \tilde{L}_s(\theta). \tag{2.115}$$

Note that Eq. (2.111) and (2.113) are valid for the range $|\phi_s - \phi_{s+1}| \leq \frac{\pi}{2}$. The space of parameters u_a satisfying these conditions is called the **minimal chamber**. If one of these phase differences crosses $\pi/2$, one has to modify the TBA equations. This phenomenon is known as the **wall-crossing** of the TBA Eq. [20]. In this section, we will study the wall-crossing of the TBA equations (2.113) and explore new TBA equations which hold on the moduli space of parameters $\{u_a\}$.

We explain the wall-crossing of the TBA equations (2.113) for the case that the turning points vary from the real value such that $\phi_2 - \phi_1$ crosses $\pi/2$ while all other phase differences $|\phi_s - \phi_{s+1}|$ are smaller than $\pi/2$. We start from the TBA Eq. (2.113) for $s = 1$ in the minimal chamber $|\phi_2 - \phi_1| < \frac{\pi}{2}$, where the integration contour is the real axis. The pole of the kernel $K_{1,2}(\theta' - \theta)$ is located at $\theta' = \theta - i\phi_1 + i\phi_2 - \frac{i\pi}{2}$ in the strip $|\text{Im}\theta| < \frac{\pi}{2}$. See Fig. 2.11. We increase the phase difference $\phi_2 - \phi_1$. If $\phi_2 - \phi_1$ crosses the value of $\pi/2$, the pole will collide with the integration contour, see Fig. 2.12a. We then deform the integration contour, which is equivalent to the real axis plus the small circle around the pole, Fig. 2.12b. We thus should pick up the contribution of the pole in the integration.

The TBA equation for $\tilde{\epsilon}_2$ also changes due to pole of the kernel $K_{2,1}(\theta' - \theta)$ at $\theta' = \theta - i\phi_2 + i\phi_1 + \frac{i\pi}{2}$. We thus find that when $\phi_2 - \phi_1$ crosses $\frac{\pi}{2}$ the TBA Eq. (2.113) are modified to

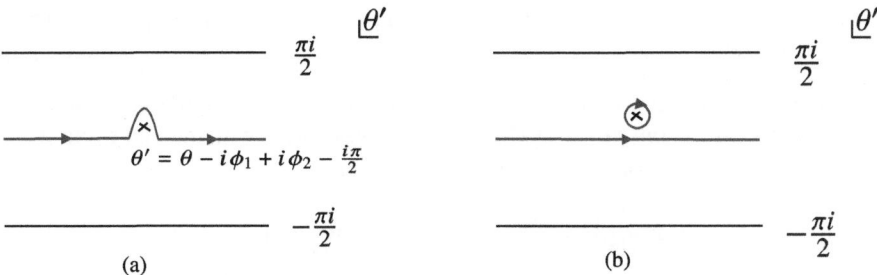

Fig. 2.12 The pole structure of the kernel of the TBA equations after $\phi_2 - \phi_1$ crossing $\pi/2$

$$\tilde{\epsilon}_1(\theta) = |m_1|e^\theta - K_{1,2} * \tilde{L}_2 - L_2(\theta - i\phi_1 - \frac{i\pi}{2}),$$

$$\tilde{\epsilon}_2(\theta) = |m_2|e^\theta - K_{2,1} * \tilde{L}_1 - K_{2,3} * \tilde{L}_3 - L_1(\theta - i\phi_2 + \frac{i\pi}{2}), \qquad (2.116)$$

$$\tilde{\epsilon}_3(\theta) = |m_3|e^\theta - K_{3,2} * \tilde{L}_2 - K_{3,4} * \tilde{L}_4,$$

$$\tilde{\epsilon}_s(\theta) = |m_s|e^\theta - K_{s,s-1} * \tilde{L}_{s-1} - K_{s,s+1} * \tilde{L}_{s+1}, \quad s = 4, \ldots, r,$$

However, these TBA equations are not closed due to the new terms on the right-hand sides of the first two equations. To obtain a closed TBA system, we need to add the two equations to the system, which are obtained by shifting the argument of the first two equations in (2.116):

$$\epsilon_1(\theta - i\phi_2 + \frac{\pi i}{2}) = i|m_1|e^{\theta + i\phi_1 - i\phi_2} - \int_{-\infty}^{\infty} d\theta' \frac{L_2(\theta' - i\phi_2)}{2\pi i \sinh(\theta - \theta')} - \tilde{L}_2(\theta),$$

$$\epsilon_2(\theta - i\phi_1 - \frac{\pi i}{2}) = -i|m_1|e^{\theta - i\phi_1 + i\phi_2} + \int_{-\infty}^{\infty} d\theta' \frac{L_2(\theta' - i\phi_1)}{2\pi i \sinh(\theta - \theta')} - \tilde{L}_1(\theta),$$

$$+ \int_{-\infty}^{\infty} d\theta' \frac{\tilde{L}_3(\theta')}{2\pi i \sinh(\theta - \theta' - i\phi_1 + i\phi_3)}.$$

$$(2.117)$$

We thus obtain a closed TBA system including Eqs. (2.116) and (2.117), which has $r + 2$ equations in total. However, the new TBA system is not directly related to the standard TBA equations of the quantum integrable model. Moreover, since the source term appears as a complex value, numerical study is not easy. It is more useful to introduce the new Y-functions

$$Y_1^n(\theta) = \frac{Y_1(\theta)}{1 + Y_2(\theta - \frac{i\pi}{2})}, \quad Y_2^n(\theta) = \frac{Y_2(\theta)}{1 + Y_1(\theta + \frac{i\pi}{2})}, \quad Y_{12}^n(\theta) = \frac{Y_1(\theta)Y_2(\theta - \frac{i\pi}{2})}{1 + Y_1(\theta) + Y_2(\theta - \frac{i\pi}{2})},$$

$$(2.118)$$

and the pseudo-energy functions $\epsilon_s^n(\theta) = -\log Y_s^n(\theta)$. We also define the shifted functions:

$$\widetilde{\epsilon}_a^{\mathrm{n}}(\theta) = -\log Y_a^{\mathrm{n}}(\theta - i\phi_s), \quad \widetilde{L}_a^{\mathrm{n}}(\theta) = \log\left(1 + Y_a^{\mathrm{n}}(\theta - i\phi_a)\right), \quad a = 1, 2, 12.$$
$$(2.119)$$

Here the mass m_{12} and its phase ϕ_{12} in the source term of $Y_{12}(\theta)$ are defined by $m_{12} = m_1 - im_2 = |m_{12}|e^{i\phi_{12}} = \Pi_{\gamma_{12}}$, which is associated with the one-cycle $\gamma_{12} = \gamma_1 + \gamma_2$. These new Y functions satisfy the relations

$$1 + Y_1(\theta) = \left(1 + Y_1^{\mathrm{n}}(\theta)\right)\left(1 + Y_{12}^{\mathrm{n}}(\theta)\right),$$
$$1 + Y_2(\theta) = \left(1 + Y_2^{\mathrm{n}}(\theta)\right)\left(1 + Y_{12}^{\mathrm{n}}(\theta + \frac{\pi i}{2})\right).$$
$$(2.120)$$

We thus can rewrite the right-hand side of the TBA equations (2.116) as the new Y-functions

$$\widetilde{\epsilon}_1^{\mathrm{n}}(\theta) = |m_1|e^\theta - K_{1,2} * \widetilde{L}_2^{\mathrm{n}} - K_{1,12}^+ * \widetilde{L}_{12}^{\mathrm{n}},$$
$$\widetilde{\epsilon}_2^{\mathrm{n}}(\theta) = |m_2|e^\theta - K_{2,1} * \widetilde{L}_1^{\mathrm{n}} - K_{2,3} * \widetilde{L}_3 - K_{2,12} * \widetilde{L}_{12}^{\mathrm{n}},$$
$$\widetilde{\epsilon}_3^{\mathrm{n}}(\theta) = |m_3|e^\theta - K_{3,2} * \widetilde{L}_2^{\mathrm{n}} - K_{3,4} * \widetilde{L}_4 - K_{2,12}^+ * \widetilde{L}_{12}^{\mathrm{n}},$$
$$\widetilde{\epsilon}_s(\theta) = |m_s|e^\theta - K_{s,s-1} * \widetilde{L}_{s-1} - K_{s,s+1} * \widetilde{L}_{s+1}, \quad s = 4, \dots, r,$$
$$(2.121)$$

where $K_{r,s}^\pm(\theta) = K_{r,s}(\theta \pm i\pi/2)$. Here we have shifted the integral contour associated with Y_{12}^{n} because the phase $|\phi_{12} - \phi_{1,2}|$ is assumed to be small enough against further wall-crossing. To derive the TBA equation for Y_{12}^{n}, we evaluate the first two TBA equations (2.116) by shifting the value of θ appropriately:

$$\log\frac{Y_1(\theta)}{1 + Y_2(\theta - \frac{\pi i}{2})} = -|m_1|e^\theta + \int \frac{d\theta'}{2\pi}\frac{\log\left(1 + Y_2(\theta' - i\phi_2)\right)}{\cosh(\theta - \theta' + i\phi_2)}$$

$$\log\frac{Y_2(\theta - \frac{\pi i}{2})}{1 + Y_1(\theta)} = -|m_2|e^{\theta - \frac{\pi i}{2}} + \int \frac{d\theta'}{2\pi}\frac{\log\left(1 + Y_1(\theta' - i\phi_1)\right)}{\cosh(\theta - \theta' - \frac{\pi i}{2} + i\phi_1)} \quad (2.122)$$

$$+ \int \frac{d\theta'}{2\pi}\frac{\log\left(1 + Y_3(\theta' - i\phi_3)\right)}{\cosh(\theta - \theta' - \frac{\pi i}{2} + i\phi_3)}.$$

After taking the sum of these two equations, the left-hand side results in

$$\log\left(\frac{Y_1(\theta)}{1 + Y_2(\theta - \frac{\pi i}{2})}\frac{Y_2(\theta - \frac{\pi i}{2})}{1 + Y_1(\theta)}\right) = \log\frac{Y_{12}^{\mathrm{n}}(\theta)}{1 + Y_{12}^{\mathrm{n}}(\theta)}. \quad (2.123)$$

The right-hand side becomes

$$
\begin{aligned}
&- |m_1|e^\theta - |m_2|e^{\theta - \frac{\pi i}{2}} + \int \frac{d\theta'}{2\pi} \frac{\log\left(1 + Y_2^n(\theta' - i\phi_2)\right)}{\cosh(\theta - \theta' + i\phi_2)} \\
&+ \int \frac{d\theta'}{2\pi} \frac{\log\left(1 + Y_1^n(\theta' - i\phi_1)\right)}{\cosh(\theta - \theta' - \frac{\pi i}{2} + i\phi_1)} + \int \frac{d\theta'}{2\pi} \frac{\log\left(1 + Y_3^n(\theta' - i\phi_3)\right)}{\cosh(\theta - \theta' - \frac{\pi i}{2} + i\phi_3)} \quad (2.124) \\
&+ \left(\int_{-\infty - i\phi_2 + \frac{\pi i}{2}}^{\infty - i\phi_2 + \frac{\pi i}{2}} d\theta' - \int_{-\infty - i\phi_1}^{\infty - i\phi_1} \frac{d\theta'}{2\pi} \right) \frac{\log\left(1 + Y_{12}^n(\theta')\right)}{2\pi i \sinh(\theta - \theta')},
\end{aligned}
$$

where we assumed ϕ_1 and $\phi_2 - \frac{\pi}{2}$ are small enough so that the shift of the integration contour does not collide with the pole of the kernel. The last integral can be evaluated by the residue of the pole, which will cancel the denominator of the left-hand side of Eq. (2.123). We then shift θ by $-i\phi_{12}$ and obtain the TBA equation for $\widetilde{\epsilon}_{12}(\theta)$:

$$
\widetilde{\epsilon}_{12}^n(\theta) = |m_{12}|e^\theta - K_{12,1}^- * \widetilde{L}_1^n - K_{12,2} * \widetilde{L}_2^n - K_{12,3}^- * \widetilde{L}_3^n. \quad (2.125)
$$

Equations (2.121) and (2.125) provide a closed TBA system with $r + 1$ equations.

Note that the original Y-function Y_s is identified to the exact WKB period Π_{γ_s} in the minimal chamber. After the wall-crossing, the new Y-functions $\{Y_{12}^n, Y_1^n, Y_2^n, Y_3, \ldots, Y_r\}$ are associated with the cycles $\{\gamma_{12} = \gamma_1 + \gamma_2, \gamma_1, \gamma_2, \ldots, \gamma_r\}$, under the identification of

$$
\begin{aligned}
-\log Y_s^n &= \frac{i}{\hbar} s[\Pi_{\gamma_s}], \quad s = 1, 2, 12, \\
-\log Y_s &= \frac{i}{\hbar} s[\Pi_{\gamma_s}], \quad s = 3, \ldots, r.
\end{aligned} \quad (2.126)
$$

By using the relations (2.120), one can rewrite the effective central charge (2.115) in terms of the Y-functions

$$
c_{\text{eff}} = \frac{6}{\pi^2} \int_{-\infty}^{\infty} d\theta\, e^\theta \left(\sum_{s=3}^{r} |m_a| \widetilde{L}_s(\theta) + |m_1| \widetilde{L}_1^n(\theta) + |m_2| \widetilde{L}_2^n(\theta) + |m_{12}| \widetilde{L}_{12}^n(\theta) \right). \quad (2.127)
$$

The effective central charge is invariant under the wall-crossing. So far we explained the first step of the wall-crossing. There arise further wall-crossings when we change the parameters of the potential. In the following part of this chapter, we will present the details of the wall-crossing of the TBA equations for a quartic potential.

2.6.1 Wall-Crossing of TBA Equation for the Quartic Oscillator

We present the wall-crossings of the TBA equations for the quartic oscillator in detail [8]. Let us consider the quartic potential

$$
V(x) - E = x^4 + u_1 x^3 + u_2 x^2 + u_3 x + u_4. \quad (2.128)
$$

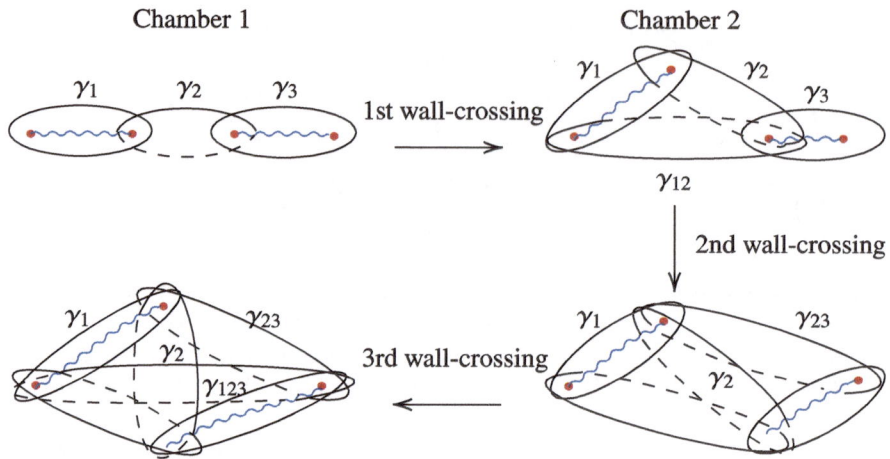

Fig. 2.13 A wall-crossing process from the minimal to the maximal chamber. One-cycles corresponding to the Y-functions are shown in each chamber. The red dots denote the turning points. The wavy lines show the branch cut on the WKB curve. The first chamber is the region of the moduli space of the potential corresponding to $|\phi_2 - \phi_1| < \frac{\pi}{2}$ and $|\phi_2 - \phi_3| < \frac{\pi}{2}$. The second chamber is determined by $\frac{\pi}{2} < \phi_2 - \phi_1 < \frac{3\pi}{2}$, $|\phi_2 - \phi_3| < \frac{\pi}{2}$ and $\phi_{12} - \phi_3 < 0$. The third one is $\frac{\pi}{2} < \phi_2 - \phi_1 < \frac{3\pi}{2}$, $\frac{\pi}{2} < \phi_2 - \phi_3 < \frac{3\pi}{2}$ and $\phi_{12} - \phi_3 < 0$. The maximal chamber is determined by $\frac{\pi}{2} < \phi_2 - \phi_1 < \frac{3\pi}{2}$, $\frac{\pi}{2} < \phi_2 - \phi_3 < \frac{3\pi}{2}$ and $\phi_{12} - \phi_3 > 0$

Suppose that the turning points change as in Fig. 2.13 by deforming the potential. We show the change of one-cycles in each process. We start from the minimal chamber where the quadratic potential with four real and different turning points, where three mass parameters $m_1, m_2,$ and m_3 associated with the cycles $\gamma_1, \gamma_2, \gamma_3$ are defined. Let ϕ_a be the phase of the mass m_a. We then deform the potential such that the phase difference $\phi_1 - \phi_2$ of masses m_1 and m_2 crosses $\pi/2$ but the difference $\phi_2 - \phi_3$ remains in the region $(-\pi/2, \pi/2)$. This process introduces a new Y-function associated with the cycle $\gamma_1 + \gamma_2$ and mass $m_1 - im_2$. Then, the wall-crossing such that $\phi_2 - \phi_3$ crosses $\pi/2$ occurs, where the Y-function associated with the new cycle $\gamma_2 + \gamma_3$ is introduced. Finally, the wall-crossing such that the difference of the phase $m_1 - im_2$ and m_3 crosses $\pi/2$ happens, which introduces the Y-function associated with the cycle $\gamma_1 + \gamma_2 + \gamma_3$. After that, the potential is deformed to the monomial potential. In the set of parameters including the monomial potential, the number of the TBA equations is maximal. We call this chamber the **maximal chamber**. We now describe the change of the TBA equations under these wall-crossings in detail.

Minimal chamber In the minimal chamber, the TBA equations are given by (2.113):

$$\tilde{\epsilon}_1 = |m_1|e^\theta - K_{1,2} * \tilde{L}_2$$
$$\tilde{\epsilon}_2 = |m_2|e^\theta - K_{2,1} * \tilde{L}_1 - K_{2,3} * \tilde{L}_3 \qquad (2.129)$$
$$\tilde{\epsilon}_3 = |m_3|e^\theta - K_{3,2} * \tilde{L}_2,$$

where $\widetilde{\epsilon}_a$ and $\widetilde{L}_a$ are defined in Eqs. (2.112).

First wall-crossing

When $\phi_2 - \phi_1$ crosses $\pi/2$, keeping $-\frac{\pi}{2} < \phi_2 - \phi_3 < \frac{\pi}{2}$, the wall-crossing in the previous section occurs. We introduce the new Y-functions

$$Y_1^{(1)}(\theta) = \frac{Y_1(\theta)}{1 + Y_2(\theta - \frac{\pi i}{2})}, \quad Y_2^{(1)} = \frac{Y_2(\theta)}{1 + Y_1(\theta + \frac{\pi i}{2})},$$

$$Y_3^{(1)}(\theta) = Y_3(\theta), \quad Y_{12}^{(1)} = \frac{Y_1(\theta)Y_2(\theta - \frac{\pi i}{2})}{1 + Y_1(\theta) + Y_2(\theta - \frac{\pi i}{2})}. \tag{2.130}$$

Let $m_a = |m_a|e^{i\phi_a}$ $(a = 1, 2, 3, 12)$ be the masses of $Y_a^{(1)}$ associated cycles γ_a where $m_{12} = m_1 - im_2$ and $\gamma_{12} = \gamma_1 + \gamma_2$. We also introduce the pseudo-energy functions $\widetilde{\epsilon}_a^{(i)}$ and $\widetilde{L}_a^{(i)}$ by

$$\widetilde{\epsilon}_a^{(i)} = -\log Y_a^{(i)}(\theta - i\phi_a), \quad \widetilde{L}_a^{(i)} = \log\left(1 + Y_a^{(i)}(\theta - i\phi_a)\right), \tag{2.131}$$

The TBA Eq. (2.121) and (2.125) are

$$\widetilde{\epsilon}_1^{(1)}(\theta) = |m_1|e^\theta - K_{1,2} * \widetilde{L}_2^{(1)} - K_{1,12}^+ * \widetilde{L}_{12}^{(1)},$$

$$\widetilde{\epsilon}_2^{(1)}(\theta) = |m_2|e^\theta - K_{2,1} * \widetilde{L}_1^{(1)} - K_{2,3} * \widetilde{L}_3^{(1)} - K_{2,12} * \widetilde{L}_{12}^{(1)},$$

$$\widetilde{\epsilon}_3^{(1)}(\theta) = |m_3|e^\theta - K_{3,2} * \widetilde{L}_2^{(1)} - K_{3,12}^+ * \widetilde{L}_{12}^{(1)},$$

$$\widetilde{\epsilon}_{12}^{(1)}(\theta) = |m_{12}|e^\theta - K_{12,1}^- * \widetilde{L}_1^{(1)} - K_{12,3}^- * \widetilde{L}_3^{(1)} - K_{12,2} * \widetilde{L}_2^{(1)}. \tag{2.132}$$

Second wall-crossing

In the second wall-crossing, $\phi_2 - \phi_3$ crosses $\pi/2$, keeping $\frac{\pi}{2} < \phi_2 - \phi_1 < \pi$, where we should pick up the contribution of the pole in the second and the third TBA equations of (2.132). This leads to the TBA equations in the third chamber

$$\widetilde{\epsilon}_1^{(2)}(\theta) = |m_1|e^\theta - K_{1,2} * \widetilde{L}_2^{(2)} - K_{1,12}^+ * \widetilde{L}_{12}^{(2)} - K_{1,32}^+ * \widetilde{L}_{32}^{(2)},$$

$$\widetilde{\epsilon}_2^{(2)}(\theta) = |m_2|e^\theta - K_{2,1} * \widetilde{L}_1^{(2)} - K_{2,3} * \widetilde{L}_3^{(2)} - K_{2,12} * \widetilde{L}_{12}^{(2)} - K_{2,32} * \widetilde{L}_{32}^{(2)},$$

$$\widetilde{\epsilon}_3^{(2)}(\theta) = |m_3|e^\theta - K_{3,2} * \widetilde{L}_2^{(2)} - K_{3,12}^+ * \widetilde{L}_{12}^{(2)} - K_{3,32}^+ * \widetilde{L}_{32}^{(2)},$$

$$\widetilde{\epsilon}_{12}^{(2)}(\theta) = |m_{12}|e^\theta - K_{12,1}^- * \widetilde{L}_1^{(2)} - K_{12,3}^- * \widetilde{L}_3^{(2)} - K_{12,2} * \widetilde{L}_2^{(2)},$$

$$\widetilde{\epsilon}_{32}^{(2)}(\theta) = |m_{32}|e^\theta - K_{32,1}^- * \widetilde{L}_1^{(2)} - K_{32,3}^- * \widetilde{L}_3^{(2)} - K_{32,2} * \widetilde{L}_2^{(2)}, \tag{2.133}$$

where the Y-functions are defined by

$$Y_2^{(2)}(\theta) = \frac{Y_2^{(1)}(\theta)}{1 + Y_3^{(1)}(\theta + \frac{\pi i}{2})}, \quad Y_3^{(2)}(\theta) = \frac{Y_3^{(1)}(\theta)}{1 + Y_2^{(1)}(\theta - \frac{\pi i}{2})},$$

$$Y_{32}^{(2)}(\theta) = \frac{Y_2^{(1)}(\theta - \frac{\pi i}{2})Y_3^{(1)}(\theta)}{1 + Y_2^{(1)}(\theta - \frac{\pi i}{2}) + Y_3^{(1)}(\theta)}, \quad Y_{\text{other}}^{(2)}(\theta) = Y_{\text{other}}^{(1)}(\theta). \tag{2.134}$$

The mass of $Y_{32}(\theta)$ is $m_{32} = m_3 - im_2$ and $\phi_{32} = \arg(m_{32})$, which is associated with the one-cycle $\gamma_{23} = \gamma_2 + \gamma_3$.

Third wall-crossing

In the third wall-crossing, $\phi_{12} - \phi_3 + \frac{\pi}{2}$ cross $\pi/2$, where the pole contributes to the TBA equations of $\epsilon_3^{(2)}$ and $\epsilon_{12}^{(2)}$. We introduce the new functions by

$$Y_3^{(3)}(\theta) = \frac{Y_3^{(2)}(\theta)}{1 + Y_{12}^{(2)}(\theta)}, \quad Y_{12}^{(3)}(\theta) = \frac{Y_{12}^{(2)}(\theta)}{1 + Y_3^{(2)}(\theta)},$$

$$Y_{123}^{(3)}(\theta) = \frac{Y_{12}^{(2)}(\theta)Y_3^{(2)}(\theta)}{1 + Y_{12}^{(2)}(\theta) + Y_3^{(2)}(\theta)}, \quad Y_{\text{other}}^{(3)}(\theta) = Y_{\text{other}}^{(3)}(\theta), \tag{2.135}$$

which leads to the TBA equations in the fourth chamber

$$\tilde{\epsilon}_1^{(3)}(\theta) = |m_1|e^\theta - K_{1,2} * \tilde{L}_2^{(3)} - K_{1,12}^+ * \tilde{L}_{12}^{(3)} - K_{1,32}^+ * \tilde{L}_{32}^{(3)} - K_{1,123}^+ * \tilde{L}_{123}^{(3)},$$

$$\tilde{\epsilon}_2^{(3)}(\theta) = |m_2|e^\theta - K_{2,1} * \tilde{L}_1^{(3)} - K_{2,3} * \tilde{L}_3^{(3)} - K_{2,12} * \tilde{L}_{12}^{(3)} - K_{2,32} * \tilde{L}_{32}^{(3)} - 2K_{2,123} * \tilde{L}_{123}^{(3)},$$

$$\tilde{\epsilon}_3^{(3)}(\theta) = |m_3|e^\theta - K_{3,2} * \tilde{L}_2^{(3)} - K_{3,12}^+ * \tilde{L}_{12}^{(3)} - K_{3,32}^+ * \tilde{L}_{32}^{(3)} - K_{3,123}^+ * \tilde{L}_{123}^{(3)},$$

$$\tilde{\epsilon}_{12}^{(3)}(\theta) = |m_{12}|e^\theta - K_{12,1}^- * \tilde{L}_1^{(3)} - K_{12,3}^- * \tilde{L}_3^{(3)} - K_{12,2} * \tilde{L}_2^{(3)} - K_{12,123}^- * \tilde{L}_{123}^{(3)},$$

$$\tilde{\epsilon}_{32}^{(3)}(\theta) = |m_{32}|e^\theta - K_{32,1}^- * \tilde{L}_1^{(3)} - K_{32,3}^- * \tilde{L}_3^{(3)} - K_{32,2} * \tilde{L}_2^{(3)} - K_{32,123}^- * \tilde{L}_{123}^{(3)},$$

$$\tilde{\epsilon}_{123}^{(3)}(\theta) = |m_{123}|e^\theta - K_{123,1}^- * \tilde{L}_1^{(3)} - K_{123,3}^- * \tilde{L}_3^{(3)} - 2K_{123,2} * \tilde{L}_2^{(3)} - K_{123,12}^+ * \tilde{L}_{12}^{(3)} - K_{123,32}^+ * \tilde{L}_{32}^{(3)}, \tag{2.136}$$

where $m_{123} = m_1 - im_2 + m_3$ and $\phi_{123} = \arg(m_{123})$ are associated with the one-cycle $\gamma_{123} = \gamma_1 + \gamma_2 + \gamma_3$. In this chamber, we have considered all the possible one-cycles between the four turning points. See Fig. 2.16. We thus conclude that the number of the TBA equations in this chamber is maximal, and call this chamber the **maximal chamber**.

We have tested the TBA equations in each chamber numerically by conforming the identification between the logarithm of the Y-function and the associating WKB periods [8].

2.6.2   TBA Equations for Monomial Potential

The TBA equations presented so far are those for a generic point in each chamber. In some special points of the moduli space of the potential, one may find some of the WKB periods appear symmetrically. The corresponding Y-functions will also be identified. The symmetry leads to a set of simpler TBA equations. A typical example is the monomial potential discussed in Chap. 1. Here, the quartic monomial potential with

$$p(x) = x^4 - E \tag{2.137}$$

is considered, where E is a positive real number. The cycles γ_1, γ_2, γ_3, γ_{12}, and γ_{32} are shown in chamber 4 in Fig. 2.13. The associated masses m_a and its phases ϕ_a satisfy the relations

$$|m_1| = |m_3| = |m_{12}| = |m_{32}| = \frac{1}{\sqrt{2}}|m_2| = \frac{1}{\sqrt{2}}|m_{123}|$$

$$\phi_1 = \phi_3 = -\frac{\pi}{4}, \quad \phi_2 = \pi, \quad \phi_{12} = \phi_{32} = \frac{\pi}{4}, \quad \phi_{123} = 0. \tag{2.138}$$

The relations (2.138) impose the symmetry on the source terms in the TBA equations (2.136). The Y-functions obey the relations

$$\tilde{\epsilon}_1(\theta) = \tilde{\epsilon}_3(\theta) = \tilde{\epsilon}_{12}(\theta) = \tilde{\epsilon}_{32}(\theta), \quad \tilde{\epsilon}_2(\theta) = \tilde{\epsilon}_{123}(\theta). \tag{2.139}$$

The six-term TBA equations (2.136) thus reduce to

$$\tilde{\epsilon}_1(\theta) = |m_1|e^\theta + \frac{1}{\pi}\int_{-\infty}^\infty d\theta' \frac{\tilde{L}_1(\theta')}{\cosh(\theta - \theta')} + \frac{\sqrt{2}}{\pi}\int_{-\infty}^\infty d\theta' \frac{\cosh(\theta - \theta')}{\cosh\left(2(\theta - \theta')\right)}\tilde{L}_2(\theta'),$$

$$\tilde{\epsilon}_2(\theta) = |m_2|e^\theta + \frac{2\sqrt{2}}{\pi}\int_{-\infty}^\infty d\theta' \frac{\cosh(\theta - \theta')}{\cosh\left(2(\theta - \theta')\right)}\tilde{L}_1(\theta') + \frac{1}{\pi}\int_{-\infty}^\infty d\theta' \frac{\tilde{L}_2(\theta')}{\cosh(\theta - \theta')}. \tag{2.140}$$

The related Y-system indicates the periodic condition $\tilde{\epsilon}_s(\theta) = \tilde{\epsilon}_s(\theta + \frac{3\pi i}{2})$. It is worth noting that these two TBA equations are precisely those derived in the quartic monomial potential [34].

For the $(r + 1)$-th order polynomial potential, there are $r(r + 1)/2$ Y-functions in the maximal chamber, which are associated with all possible cycles between $r + 1$ turning points. This provides a closed system with $r(r + 1)/2$ TBA equations. For the monomial potential in the maximal chamber, the symmetry reduces the $r(r + 1)/2$-TBA to the A_r-type TBA Eq. [34].

For the monomial potential, it is natural to use the effective central charge of the reduced TBA equations

$$c_{\text{eff}}^{\text{mon}} = \frac{1}{r + 1}c_{\text{eff}} = \frac{r}{r + 3}. \tag{2.141}$$

For even $r = 2n$, the effective central charge (2.141) coincides with that of the non-unitary minimal model $M_{2,2n+3}$ whose central charge $c = 1 - 3\frac{(n+1)^2}{n+3}$ and the lowest conformal dimension $\Delta_{\min} = \frac{1-(n+1)^2}{8(n+3)}$. The effective central charge is defined by $c_{\text{eff}} = c - 24\Delta_{\min}$. The same reduced TBA equations at the monomial point have also been derived from the minimal model $M_{2,2n+3}$ [35]. See also the Appendix for a brief review.

2.7   Voros Spectrum for Double Well Potential

For the double well potential (2.142) where the potential is invariant under $x \to -x$, there are only two chambers, the minimal chamber and the maximal chamber.

In this section, we apply the TBA equations by solving the spectral problem of the Schrödinger equation (2.1) with the double well potential. We first test our TBA equations by comparing them with the WKB periods in the minimal chamber and also the maximal chamber. We will also combine the TBA equations with the exact quantization condition to compute the Voros spectrum.

We consider the double-well potential

$$V(x) = 2(\frac{x^4}{2} - \frac{x^2}{4} + \frac{1}{32}). \tag{2.142}$$

The structure of turning points changes when E varies. When $0 < E < \frac{1}{16}$, there are four real turning points, where the system is in the minimal chamber. For $E > \frac{1}{16}$ there are two real and two pure complex turning points. For $E < 0$, there are four complex turning points. These two regions of E correspond to the maximal chamber. See Fig. 2.14 .

2.7.1   Minimal Chamber

We first consider $0 < E < \frac{1}{16}$, where four turning points $\{-a, -b, b, a\}$ are all real, where $a = \sqrt{\frac{1}{4} + \sqrt{E}}$ and $b = \sqrt{\frac{1}{4} - \sqrt{E}}$. We denote the one cycles going around the intervals $(-a, -b), (-b, b), (b, a)$ as $\gamma_1, \gamma_2, \gamma_3$ respectively (Fig. 2.15). The

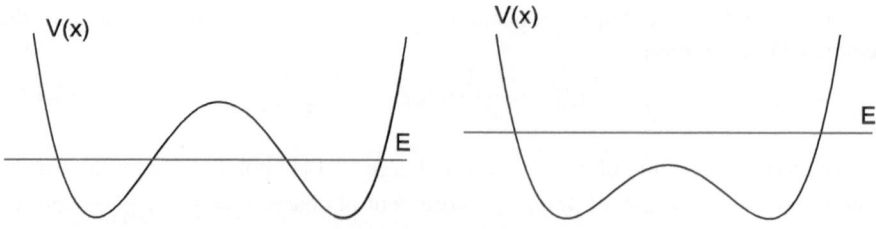

Fig. 2.14 The double well oscillator for $V(x) = 2(\frac{x^4}{2} - \frac{x^2}{4} + \frac{1}{32})$ with $0 < E < \frac{1}{16}$ (left) and $E > \frac{1}{16}$ (right)

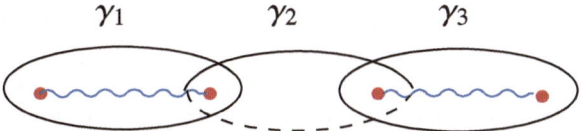

γ_1 $\qquad$ γ_2 $\qquad$ γ_3

Fig. 2.15 The period structure of the double well oscillator with $V(x) = 2(\frac{x^4}{2} - \frac{x^2}{4} + \frac{1}{32})$ and $0 < E < \frac{1}{16}$. The turning points are denoted by the red dots

Table 2.1 The quantum correction of the WKB period for the quartic potential $V(x) = 2(\frac{x^4}{2} - \frac{x^2}{4} + \frac{1}{32})$ with energy $\frac{2}{50}$, $y^2 = x^4 - \frac{x^2}{2} + \frac{9}{400}$, which is computed by using the differential operator acting on the classical period presented in Sect. 2.1. The TBA equations are solved by using the Fourier discretization with 2^{14} points and cutoff $(-40, 40)$

n	$\Pi_{\gamma_1}^{(n)}$	$m_1^{(n)}$	$i\Pi_{\gamma_1}^{(n)}$	$m_2^{(n)}$
0	0.1359565096829	0.1359565096829	0.103887649235	0.103887649235
1	3.580623900785	−3.580623900783	−4.966403984067	−4.966403984064
2	642.47945147513	642.47945147506	555.31920041992	555.31920041986

masses and the classical periods defined in (2.59) can be calculated by using the elliptic integrals[7]

$$m_1 = \frac{2a^3}{3}\left((1 + k'^2)E(k) - 2k'^2 K(k)\right),$$
$$m_2 = \frac{4a^3}{3}\left((1 + k'^2)E(k') - k^2 K(k')\right),$$

(2.144)

where $k^2 = 1 - b^2/a^2$ and $k'^2 = 1 - k^2$ are the elliptic moduli. Since the potential is an even function of x, $\Pi_{\gamma_1} = \Pi_{\gamma_3}$ which leads to $m_1 = m_3$. The TBA equations are

$$\epsilon_1(\theta) = m_1 e^\theta - K * L_2, \quad \epsilon_2(\theta) = m_2 e^\theta - K * L_1 - K * L_3, \quad \epsilon_3(\theta) = m_3 e^\theta - K * L_2,$$

(2.145)

where the kernel $K(\theta)$ is defined in Eq. (2.95). Since $m_1 = m_3$, one finds $\epsilon_1 = \epsilon_3$ and $L_1 = L_3$. One can solve the TBA equations numerically to obtain the pseudo-energies ϵ_s. Then the coefficient $m_s^{(n)}$ in (2.101) is calculated numerically. On the other hand, the quantum correction of the WKB periods can be computed by using the differential operator [8] which is also reviewed in Sect. 2.1. In Table 2.1, we compare the quantum corrections of WKB periods and the expansion of the TBA equations. The numerical results test the identification (2.102) with very high precision.

---

[7] The first- and second-kind elliptic integrals are defined by

$$E(m) = \int_0^{\frac{\pi}{2}} \sqrt{1 - m^2 \sin^2\theta}\, d\theta, \quad K(m) = \int_0^{\frac{\pi}{2}} \frac{d\theta}{\sqrt{1 - m^2 \sin^2\theta}}.$$

(2.143)

So far, we have shown that our TBA equations can reproduce the asymptotics of the WKB periods. As we have seen in Eq. (2.68), our TBA equations can also provide the discontinuity of the WKB periods. We thus conclude that our TBA equations describe the exact WKB periods under the identification (2.62). We then combine our TBA equations with the exact quantization condition and compute the Voros spectrum with given energy E. This provides also another non-perturbative test of our TBA equations.

We now write down the exact quantization condition for the exact WKB periods. There is an ambiguity in defining the Borel non-summable WKB periods associated with the classically allowed region, which can be resolved by introducing the median resummation of the WKB periods. To do this, we first express the Borel resummation of the WKB periods by using our TBA equations (2.145):

$$\frac{i}{\hbar} s_{\mp}[\Pi_{\gamma_1}] = i m_1 e^{\theta \pm i 0_+} - \int_{-\infty}^{\infty} \frac{d\theta'}{2\pi i} \frac{L_2(\theta')}{\sinh(\theta - \theta' \pm i 0_+)},$$
$$\frac{i}{\hbar} s[\Pi_{\gamma_2}] = m_2 e^{\theta} - 2K * L_1,$$
(2.146)

where we have used the identification (2.62). Note that the integration in the first equation is singular at $\theta' = \theta$. We use two prescriptions of $\pm i 0_+$ in Eqs. (2.146) that correspond to the two lateral Borel transformations. We then introduce the "median" resummation [6, 7][8]: $s_{\text{med}}[\Pi_{\gamma_1}]$ defined by the average of two lateral resummation $s_{\text{med}}[\Pi_{\gamma_1}] = \frac{1}{2}(s_+[\Pi_{\gamma_1}] + s_-[\Pi_{\gamma_1}])$. Rewriting the WKB periods in terms of TBA equations, one obtains

$$\frac{1}{\hbar} s_{\text{med}}[\Pi_{\gamma_1}] = m_1 e^{\theta} + \mathrm{P} \int_{-\infty}^{\infty} \frac{d\theta'}{2\pi} \frac{L_2(\theta')}{\sinh(\theta - \theta')},$$
$$\frac{i}{\hbar} s[\Pi_{\gamma_2}] = m_2 e^{\theta} - 2K * L_1,$$
(2.147)

where P denotes the principal value of the singular integral.

The exact quantization condition for potential (2.142) has been derived in [6, 14]

$$\cos\left(\frac{1}{\hbar} s_{\text{med}}[\Pi_{\gamma_1}](\hbar)\right) + \frac{1}{\sqrt{1 + e^{-i s[\Pi_{\gamma_2}(\hbar)]}}} = 0,$$
(2.148)

or equivalently

$$\mathcal{J}_{\epsilon}(\hbar) = \frac{1}{\hbar} s_{\text{med}}[\Pi_{\gamma_1}](\hbar) + \epsilon \tan^{-1}\left(e^{-\frac{i}{2\hbar} s[\Pi_{\gamma_2}](\hbar)}\right) = 2\pi\left(k + \frac{1}{2}\right), \quad k \in \mathbb{Z}_{\geq 0},$$
(2.149)

where $\epsilon = \pm 1$ represents the parity. The n-th excited state of the spectrum corresponds to $n = 2k - (\epsilon - 1)/2$.

---

[8] See also [36, 37].

Table 2.2 The first column shows the Voros spectrum x_n for potential (2.142) and energy $E/2 = 1/50$, which are computed by using TBA equations and the exact quantization condition. The TBA equations are solved by using a discretization with 2^{18} points and $L = 35$. The second column presents the energy $E_n(\hbar)$ with the corresponding Planck constant $\hbar = 1/x_n$

Level n	x_n	$\frac{1}{2}E_n(x_n^{-1})$
0	18.7282060	0.0199999237
1	24.1286618	0.0199999365
2	68.7046550	0.0199999995
3	69.1321599	0.0199999995

Now let us combine our TBA equations with the exact quantization condition (2.149) to compute the Voros spectrum $x(E) = 1/\hbar$ introduced in Sect. 2.2 with given energy E. Starting from the Planck constant of these values of x, we can test the accuracy of our method by calculating the energy $E(x^{-1})$, which can be computed by other standard methods. In Table 2.2, we present the Voros spectrum x_n with level n computed by using our method and the energy $E(x_n^{-1})$ computed by the standard Rayleigh–Ritz method with complex dilatation [38]. Given the Voros spectrum computed by using the TBA system, the energy produces the value $E/2 = 1/50$, which is correct in the order 10^{-7} for $n = 0, 1$ and 10^{-9} for higher n.

So far, we have shown the numeric test of the TBA equations in the minimal chamber with a particular example of the double well potential. We have also tested our TBA equations with other generic points in the minimal chamber. We also find the TBA Eq. (2.145) are not valid anymore when the phase difference $|\phi_s - \phi_{s+1}|$ crosses $\pi/2$.

As we mentioned in Sect. 2.3, the effective central charge imposes a strong constraint on the classical WKB periods and their quantum correction. For the double well potential, one finds $m_1 = m_3$ and $m_1^{(1)} = m_3^{(1)}$. The effective central charge leads to

$$c_{\text{eff}} = -\frac{6}{\pi}(m_1 m_2^{(1)} + m_2 m_1^{(1)}), \tag{2.150}$$

which leads to

$$\frac{i\pi}{3} = \Pi_{\gamma_1}^{(0)}\Pi_{\gamma_2}^{(1)} - \Pi_{\gamma_2}^{(0)}\Pi_{\gamma_1}^{(1)}. \tag{2.151}$$

One can confirm this relation by using the numeric data, e.g. Table 2.1.

2.7.2 Maximal Chamber

We now increase the energy E above the value of $\frac{1}{16}$ such that there are only two turning points on the real axis. See the right of Fig. 2.14. In this case, the system

Fig. 2.16 The period structure of the double well oscillator with $V(x) = 2(\frac{x^4}{2} - \frac{x^2}{4} + \frac{1}{32})$ and $E > \frac{1}{16}$. The turning points are denoted by the red dots

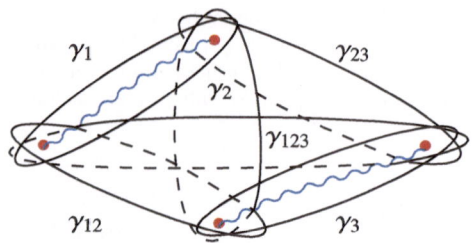

moves to the maximal chamber of the moduli space, where the structure of the periods is shown in Fig. 2.16. The TBA equations are given by (2.136).

The masses are related by

$$m_1 = m_3, \quad m_{12} = m_{32}, \tag{2.152}$$

because of the parity symmetry $x \to -x$. The phases of the masses are

$$\phi_1 = -\phi_{12} = -\alpha, \quad \phi_2 = \pi, \quad \phi_{123} = 0, \tag{2.153}$$

where α depends on the energy E. We thus find the identification between the Y-functions

$$\tilde{\epsilon}_1 = \tilde{\epsilon}_3, \quad \tilde{\epsilon}_{12} = \tilde{\epsilon}_{32}. \tag{2.154}$$

The TBA equations (2.136) thus reduce to

$$
\begin{aligned}
\tilde{\epsilon}_1(\theta) &= |m_1|e^{\theta} - K_{1,2} * \tilde{L}_2 - 2K_{1,12}^+ * \tilde{L}_{12} - K_{1,123}^+ * \tilde{L}_{123}, \\
\tilde{\epsilon}_2(\theta) &= |m_2|e^{\theta} - 2K_{2,1} * \tilde{L}_1 - 2K_{2,12} * \tilde{L}_{12} - 2K_{2,123} * \tilde{L}_{123}, \\
\tilde{\epsilon}_{12}(\theta) &= |m_{12}|e^{\theta} - 2K_{12,1}^- * \tilde{L}_1 - K_{12,2} * \tilde{L}_2 - K_{12,123}^- * \tilde{L}_{123}, \\
\tilde{\epsilon}_{123}(\theta) &= |m_{123}|e^{\theta} - 2K_{123,1}^- * \tilde{L}_1 - 2K_{123,2} * \tilde{L}_2 - 2K_{123,12}^+ * \tilde{L}_{12}.
\end{aligned}
\tag{2.155}
$$

In Table 2.3, we test the TBA equations by comparing the quantum correction of the WKB periods with the expansion of the logarithm of the corresponding Y-functions and find a good agreement.

In Fig. 2.16, the cycle γ_{123} is associated with the two turning points on the real axis. Since the corresponding WKB period is Borel non-summable, we use the median resummation

$$
\begin{aligned}
\frac{1}{\hbar} s_{\mathrm{med}} \left(\Pi_{\gamma_{123}} \right)(\hbar) = {}& m_{123}e^{\theta} + 2i \int_{-\infty}^{\infty} \frac{d\theta'}{2\pi} \left(\frac{\tilde{L}_1(\theta')}{\cosh(\theta - \theta' - i\alpha)} - \frac{\tilde{L}_{12}(\theta')}{\cosh(\theta - \theta' + i\alpha)} \right) \\
& - 2P \int_{-\infty}^{\infty} \frac{d\theta'}{2\pi} \frac{\tilde{L}_2(\theta')}{\sinh(\theta - \theta')}.
\end{aligned}
\tag{2.156}
$$

Table 2.3 The quantum correction of the WKB period for the quartic potential $V(x) = 2(\frac{x^4}{2} - \frac{x^2}{4} + \frac{1}{32})$ with energy $E/2 = \frac{1}{2}$, $y^2 = x^4 - \frac{x^2}{2} + \frac{15}{16}$, which is computed by using the differential operator acting on the classical period presented in Sect. 2.1. The TBA equations (2.155) are solved by using the Fourier discretization with 2^{18} points and cutoff $(-35, 35)$

n	$\Pi_{\gamma_1}^{(n)}$	$m_1^{(n)}$
0	$2.003950934674 - 1.409640119378i$	$2.003950934674 - 1.409640119378i$
1	$-0.1380337977730 - 0.1641860540267i$	$0.1380337977754 + 0.1641860540247i$
2	$0.04726770527143613 - 0.009695344018921312i$	$0.04726770527126 - 0.009695344401915i$

n	$i\Pi_{\gamma_2}^{(n)}$	$m_2^{(n)}$	$\Pi_{\gamma_{123}}^{(n)}$	$m_{123}^{(n)}$
0	-2.819280238755	-2.819280238755	4.007901869348	4.007901869348
1	-0.3283721080535	-0.3283721080494	-0.2760675955462	0.2760675955508
2	-0.019390688037842604	-0.01939068803831	0.09453541054287	0.09453541054252

n	$\Pi_{\gamma_{12}}^{(n)}$	$m_{12}^{(n)}$
0	$2.003950934674 + 1.409640119378i$	$2.003950934674 + 1.409640119378i$
1	$-0.1380337977731 + 0.1641860540267i$	$0.1380337977754 - 0.1641860540247i$
2	$0.04726770527144 + 0.00969534401892i$	$0.04726770527126 + 0.00969534401915i$

Table 2.4 The first column shows the Voros spectrum x_n for potential (2.142) and energy $E/2 = 1/2$, i.e. $p(x) = x^4 - \frac{x^2}{2} - \frac{15}{16}$, which are computed by using TBA equations and the exact quantization condition. The TBA equations are solved by using a discretization with 2^{18} points and $L = 35$. The second column presents the energy $E_n(\hbar)$ with the corresponding Planck constant $\hbar = 1/x_n$

Level n	x_n	$\frac{1}{2}E_n(x_n^{-1})$
0	0.9508897549193	0.500001544
1	2.362904672569	0.500000048
2	3.938258727761	0.500000001
3	5.499130925612	0.500000000

The exact quantization condition in this case is obtained by the analytic continuation of the one in the minimal chamber

$$\frac{1}{\hbar}s_{\text{med}}\big[\Pi_{\gamma_{123}}\big](\hbar) - 2(-1)^n \tan^{-1}\left(e^{-\frac{i}{2\hbar}s[\Pi_{\gamma_2}](\hbar)}\right) = 2\pi\left(n + \frac{1}{2}\right), \quad n \in \mathbb{Z}_{\geq 0}, \tag{2.157}$$

where the contribution from the periods Π_{γ_2} for the cycle γ_2 between the complex turning points is associated with the complex instanton.

In Table 2.4, we present the Voros spectrum x_n computed by using our method for the potential (2.142) with energy $E/2 = 1/2$ and the energy $E(x_n^{-1})$ computed by the standard Rayleigh–Ritz method with complex dilatation. Given the Voros spectrum computed by using the TBA system, the energy produces the value $E/2 = 1/2$.

So far, we have computed the Voros spectrum with given energy by using our TBA equations and the exact quantization condition. We can also solve the energy spectrum conversely from the Voros spectrum by using the energy and the Planck constant as in Fig. 2.7.

2.8   Wall-Crossing and Fock–Goncharov Coordinates

The Voros symbol which is defined by the exponential of the Borel-resummed WKB periods is also represented as the cross-ratio of the Wronskians of the subdominant solutions of the ODE. The cross-ratio is also regarded as the Fock–Goncharov coordinates of the moduli space of the Hitchin system [39]. In the conformal limit, the Hitchin system reduces to the ODE. The Fock–Goncharov coordinates are constructed from the Stokes graph of the WKB curve of the ODE [12]. This gives another way of understanding the discontinuity formula and the new Y-functions introduced in the wall-crossing of the TBA equations.

2.8.1   WKB Triangulation and Fock-Goncharov Coordinates

Let us consider the second–order ODE (2.2), where $p(x)$ is a polynomial of order $r + 1$ in x with simple zeros. $x = \infty$ is an irregular singular point of the ODE. The Stokes lines emerging from this irregular singular point will divide the complex plane into $r + 3$ sectors. In each sector, there is a unique subdominant solution denoted $\hat{y}_k$ in (2.85). It is thus convenient to represent the irregular singular point as $r + 3$ marked points $Q_k, k = 0, \ldots, r + 2$, at the infinity of each sector. We can draw the Stokes graph of the Stokes lines as we explained in Sect. 2.2.

We discuss the Stokes graph according to the change of the phase of $\hbar$. In Fig. 2.17, we show the Stokes graphs for the case of $p(x) = x^2 - 1$ with different phases of $\hbar$. The topology of the Stokes graph changes when the phase of $\hbar$ crosses $\pi/2$. This transition of Stokes graph is called the **flip**.

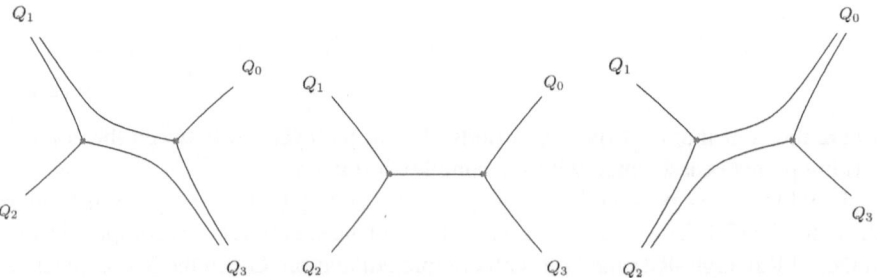

Fig. 2.17 The Stokes graph for the case $p(x) = x^2 - 1$. The phase of $\text{Arg}(\hbar)$ is fixed to be $2\pi/5, \pi/2, 3\pi/5$. Here the cross point labels the turning point of $p(x)$

At $\arg(\hbar) = \frac{\pi}{2}$, the finite Stokes line, which has both ends at zeros, appears. In the following, we will not treat the Stokes graph which contains the finite Stokes line.

In each Stokes graph, the Stokes lines divide the complex plane into several cells. In each cell, the lines (2.40) whose starting point is an inner point of the cell, define a foliation of the cell. They are homotopically equivalent. Choosing a representative curve in each cell, one obtains the graph made of the representative curves and the boundary edges, which is called the **WKB triangulation** T_{WKB}. Here the boundary edges are the lines connecting two neighboring marked points at infinity. In Fig. 2.18, we show the WKB triangulation for the first and third Stokes graphs in Fig. 2.17. Here the black lines are the boundary edges of the triangulation. The red line denotes the representative curve in the cell.

Two triangles with a common edge E make up a quadrilateral Q_E. Let us consider a quadrilateral Q_E with vertices Q_1, Q_2, Q_3, Q_4 in anticlockwise order. Here the common edge is denoted by $E(Q_1, Q_3)$, which connects vertex Q_1 and Q_3. The Fock–Goncharov coordinate associating with this edge is defined by [39]

$$X_{E(Q_1, Q_3)} = -\frac{(\hat{y}_1 \wedge \hat{y}_2)(\hat{y}_3 \wedge \hat{y}_4)}{(\hat{y}_2 \wedge \hat{y}_3)(\hat{y}_4 \wedge \hat{y}_1)}. \tag{2.158}$$

The wedge product here is the Wronskian defined in (2.86):

$$\hat{y}_{k_1} \wedge \hat{y}_{k_2} = \hbar^{\frac{2}{r+3}}\left(\hat{y}_{k_1}(x, u_a, \hbar)\partial_x \hat{y}_{k_2}(x, u_a, \hbar) - \hat{y}_{k_2}(x, u_a, \hbar)\partial_x \hat{y}_{k_1}(x, u_a, \hbar)\right). \tag{2.159}$$

We then consider the case in Fig. 2.18, where flip $E(Q_1, Q_3) \to E(Q_0, Q_2)$ occurs. Before the flip, the Fock–Goncharov coordinate for the quadrilatelal $Q_{E(Q_1, Q_3)}$ is

$$X_{E(Q_1, Q_3)} = -\frac{(\hat{y}_1 \wedge \hat{y}_2)(\hat{y}_3 \wedge \hat{y}_0)}{(\hat{y}_2 \wedge \hat{y}_3)(\hat{y}_0 \wedge \hat{y}_1)}. \tag{2.160}$$

After the flip, one has the quadrilateral $Q_{E(Q_0, Q_2)}$, whose coordinate is given by

$$X_{E(Q_0, Q_2)} = -\frac{(\hat{y}_0 \wedge \hat{y}_1)(\hat{y}_2 \wedge \hat{y}_3)}{(\hat{y}_1 \wedge \hat{y}_2)(\hat{y}_3 \wedge \hat{y}_0)}. \tag{2.161}$$

The Fock–Goncharov coordinates in the process of a flip are related by

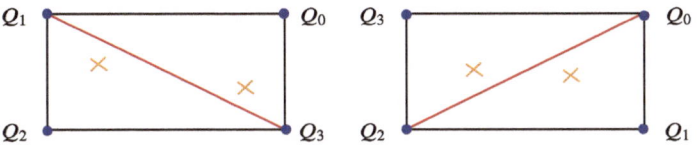

Fig. 2.18 The WKB triangulation for the case $p(x) = x^2 - 1$. The phase of $\text{Arg}(\hbar)$ is fixed to be $2\pi/5$, $3\pi/5$

$$\chi_{E(Q_0, Q_2)} = \frac{1}{\chi_{E(Q_1, Q_3)}}.$$ (2.162)

More general flips of the Stokes graph for various gauge theories can be found in [39]. As we will see later, the Y-functions are related to the Fock–Goncharov coordinates.

In the following, we will illustrate the wall-crossing of the Schrödinger equation with the cubic potential, by using the WKB triangulation. We will discuss the relation to the Y-functions and their discontinuity.

2.8.2   Wall-Crossing of Cubic Oscillator

We represent the WKB curve of the cubic oscillator as:

$$y^2 = x^3 - u_2 x - u_3.$$ (2.163)

We change the parameter u_2 for fixed u_3, which corresponds to the energy of the Schrödinger equation. There is a wall that divides the complex u_2-plane into two chambers, the minimal and the maximal chamber. It is called the **marginal stability wall**. See Fig. 2.19 for its shape in the case of $u_3 = 1/8$. The minimal chamber is the region outside of the wall, which contains the two types of BPS (Bogomol'nyi–Prasad–Sommerfield) states associated with the cycles γ_1 and γ_2. The maximal chamber contains the three types of BPS states associated with the cycles γ_1, γ_2, and γ_{12}. We will discuss the Stokes graphs and their flips using this example.

Fig. 2.19 The marginal stability wall on the complex u_2 plane for $u_3 = 1/8$

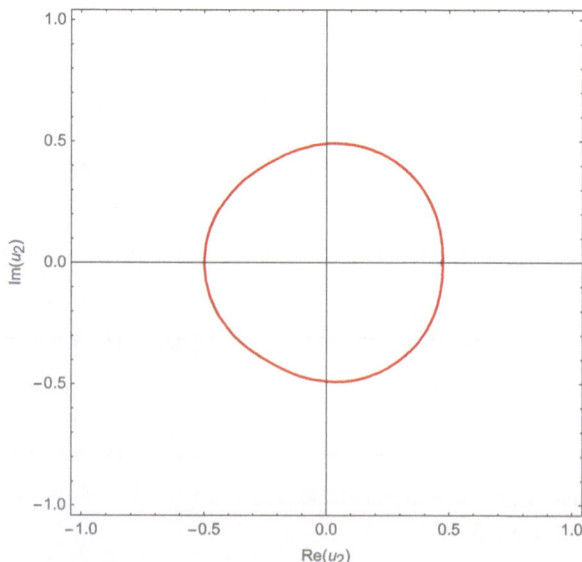

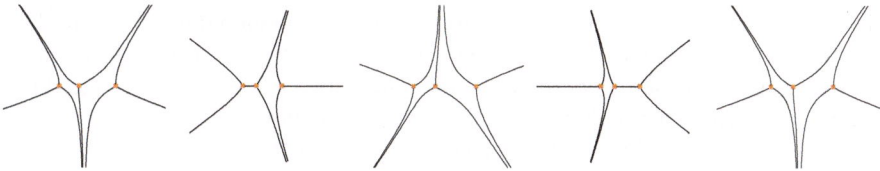

Fig. 2.20 The Stokes graph of the polynomial $p(x) = x^3 - \frac{3}{4}x - \frac{1}{8}$. Here we have fixed the phase of $\hbar$ to be $-\pi/4, 0, \pi/4, \pi/2, 3\pi/4$. The orange crosses are the zeros of $p(x)$

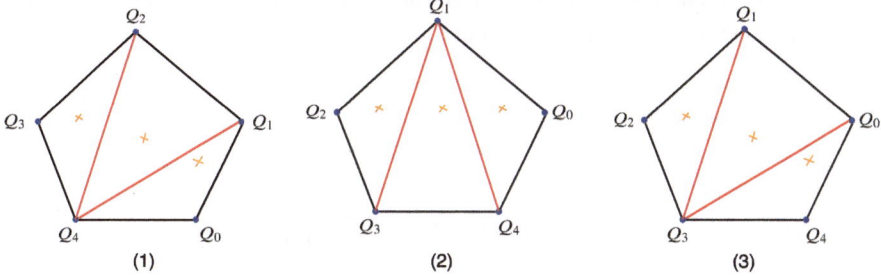

Fig. 2.21 The WKB triangulation of the polynomial $p(x) = x^3 - \frac{3}{4}x - \frac{1}{8}$, where we have fixed the phase of $\hbar$ to be $-\pi/4, \pi/4, 3\pi/4$

Let us start with the region outside the marginal stability wall, whose typical Stokes graph and WKB triangulation are shown in Figs. 2.20 and 2.21, respectively.

As we can see, the flips occur at $\arg(\hbar) = 0, \pi/2$, and the first and the fifth graphs have the same topology.

Let us consider the Fock–Goncharov coordinates (2.158) of the second WKB triangulation in Fig. 2.21 for example,

$$\mathcal{X}^{(2)}_{E(Q_1,Q_4)} = -\frac{(\hat{y}_1 \wedge \hat{y}_3)(\hat{y}_4 \wedge \hat{y}_0)}{(\hat{y}_3 \wedge \hat{y}_4)(\hat{y}_0 \wedge \hat{y}_1)}, \quad \mathcal{X}^{(2)}_{E(Q_1,Q_3)} = -\frac{(\hat{y}_1 \wedge \hat{y}_2)(\hat{y}_3 \wedge \hat{y}_4)}{(\hat{y}_2 \wedge \hat{y}_3)(\hat{y}_4 \wedge \hat{y}_1)}. \quad (2.164)$$

Moreover, the (relabelled-) Y-functions Y_1 and Y_2 in (2.87) are defined by

$$Y_2(e^{-\frac{\pi i}{2}}\hbar, u_a) = \frac{(\hat{y}_{-2} \wedge \hat{y}_1)(\hat{y}_{-1} \wedge \hat{y}_0)}{(\hat{y}_{-2} \wedge \hat{y}_{-1})(\hat{y}_0 \wedge \hat{y}_1)}, \quad \frac{1}{Y_1(\hbar, u_a)} = \frac{(\hat{y}_1 \wedge \hat{y}_2)(\hat{y}_{-2} \wedge \hat{y}_{-1})}{(\hat{y}_{-2} \wedge \hat{y}_2)(\hat{y}_{-1} \wedge \hat{y}_1)}. \quad (2.165)$$

Note that we have used the relabelled Y-functions $Y_k \to Y_{r+1-k}$ in (2.87). Also $\hat{y}_k \propto \hat{y}_{k+5}$, since there is no non-trivial monodromy around the origin of the x-plane. One thus finds that the Fock–Goncharov coordinates are related to the Y-functions by

$$\mathcal{X}^{(2)}_{E(Q_1,Q_4)} = Y_2(e^{-\frac{\pi i}{2}}\hbar), \quad \mathcal{X}^{(2)}_{E(Q_1,Q_3)} = \frac{1}{Y_1(\hbar)}. \quad (2.166)$$

These Fock–Goncharov coordinates of this WKB triangulation thus satisfy the TBA equations in the minimal chamber (2.99) with $r = 2$. Note that this identification between Fock–Goncharov coordinates and Y-functions is correct in the period $0 < \text{Arg}(\hbar) < \frac{\pi}{2}$, where the Stokes graph has the same topology.

As we have seen in Fig. 2.21, flip $E(Q_2, Q_4) \to E(Q_1, Q_3)$ and $E(Q_1, Q_4) \to E(Q_0, Q_3)$ occur at $\text{Arg}(1/\hbar) = 0$ and $\text{Arg}(\hbar) = \pi/2$ respectively, where the Fock–Goncharov coordinates will be transformed. It is thus interesting to see how the related Y-functions change in this progress. We then consider the Fock–Goncharov coordinates of the first and the third WKB triangulation in Fig. 2.21 denoted by $\mathcal{X}_E^{(1)}$ and $\mathcal{X}_E^{(3)}$ respectively. They are related to the Y-functions by

$$\mathcal{X}_{E(Q_1, Q_4)}^{(1)} = \frac{Y_2(e^{-\frac{\pi i}{2}}\hbar)}{1 + Y_1(\hbar)}, \quad \mathcal{X}_{E(Q_2, Q_4)}^{(1)} = Y_1(\hbar) \tag{2.167}$$

$$\mathcal{X}_{E(Q_0, Q_3)}^{(3)} = \frac{1}{Y_2(e^{-\frac{\pi i}{2}}\hbar)}, \quad \mathcal{X}_{E(Q_1, Q_3)}^{(3)} = \left(\frac{Y_1(\hbar)}{1 + Y_2(e^{-\frac{\pi i}{2}}\hbar)}\right)^{-1}. \tag{2.168}$$

Comparing with the Fock–Goncharov coordinates of the second WKB triangulation, we can find out how the Y-functions change during the flip. Recall the identification between the Y-function and WKB period in the minimal chamber (2.62),

$$\epsilon_1(\theta + \frac{\pi i}{2}) = \frac{i}{\hbar}s(\Pi_{\gamma_1})(\hbar), \quad \epsilon_2(\theta) = \frac{i}{\hbar}s(\Pi_{\gamma_2})(\hbar). \tag{2.169}$$

Note that $\mathcal{X}_{E(Q_1, Q_4)}^{(2)}$ $(\mathcal{X}_{E(Q_1, Q_4)}^{(1)})$ denotes the coordinates soon before (after) the flip at positive real $\hbar$. The transformation of the Y-functions from (2.166) to (2.167) during the flip can reproduce the discontinuity formula (2.60)[9]

$$\text{disc}_{-\frac{\pi}{2}}(\Pi_{\gamma_2})(\hbar) = i\hbar \log\left(1 + s\left(e^{-\frac{i}{\hbar}\Pi_{\gamma_1}}\right)\right). \tag{2.170}$$

We then consider the flip at the pure imaginary $\hbar$. The transformation of the Y-functions from (2.166) to (2.168) during the flip leads to the discontinuity formula

$$\text{disc}_0(\Pi_{\gamma_2})(\hbar) = -i\hbar \log\left(1 + s\left(e^{-\frac{i}{\hbar}\Pi_{\gamma_1}}\right)\right), \tag{2.171}$$

which is consistent with the discontinuity formula (2.61).

We then move to the region inside the marginal stability wall, whose Stokes graph and WKB triangulation are shown in Figs. 2.22 and 2.23. The three flips $E(Q_1, Q_4) \to E(Q_0, Q_2)$, $E(Q_2, Q_4) \to E(Q_0, Q_3)$ and $E(Q_0, Q_2) \to E(Q_1, Q_3)$ occur at certain phase in the period $(\pi/4, \pi/2)$, $(\pi/2, 3\pi/4)$ and $(3\pi/4, 5\pi/4)$ respectively. The Fock–Goncharov coordinates for the second WKB triangulation in Fig. 2.23 are

---

[9] Here we have used $Y_{1,2}(-\hbar) = -1/Y_{1,2}(\hbar)$. See also [40, 41].

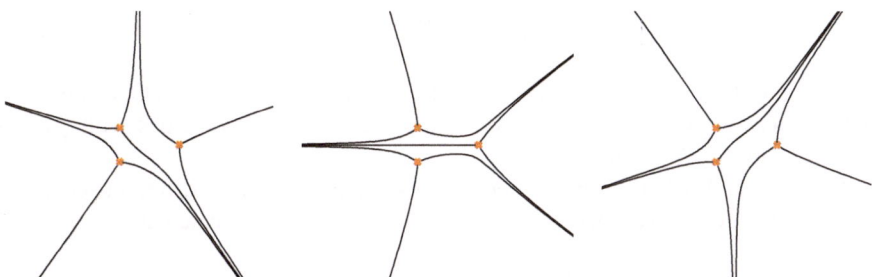

Fig. 2.22 The Stokes graph of the polynomial $p(x) = x^3 - \frac{1}{4}x - \frac{1}{8}$. Here we have fixed the phase of $\hbar$ to be $\pi/4, \pi/2, 3\pi/4$. The yellow crosses are the zeros of $p(x)$. The Stokes graphs appear periodically with the period π

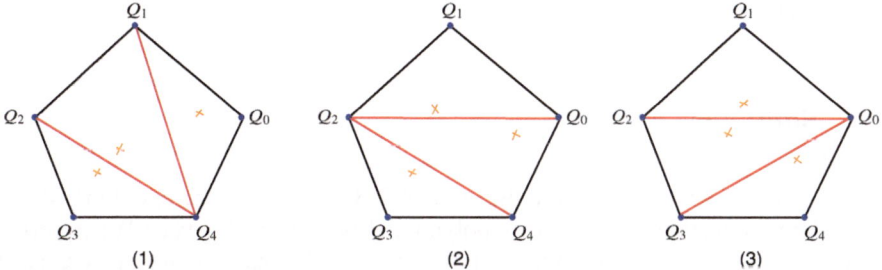

Fig. 2.23 The WKB triangulation of the polynomial $p(x) = x^3 - \frac{1}{4}x - \frac{1}{8}$, where we have fixed the phase of $\hbar$ to be $\pi/4, \pi/2, 3\pi/4$

$$\begin{aligned}
\mathcal{X}^{(2),\text{in}}_{E(Q_0,Q_2)} &= -\frac{(\hat{y}_0 \wedge \hat{y}_1)(\hat{y}_2 \wedge \hat{y}_4)}{(\hat{y}_1 \wedge \hat{y}_2)(\hat{y}_4 \wedge \hat{y}_0)} = Y_2^{\text{n}}(e^{-\frac{\pi i}{2}}\hbar)^{-1} \\
\mathcal{X}^{(2),\text{in}}_{E(Q_2,Q_4)} &= -\frac{(\hat{y}_2 \wedge \hat{y}_3)(\hat{y}_4 \wedge \hat{y}_0)}{(\hat{y}_3 \wedge \hat{y}_4)(\hat{y}_0 \wedge \hat{y}_2)} = Y_{12}^{\text{n}}(\hbar),
\end{aligned} \tag{2.172}$$

where Y^{n} is the Y-function after the wall-crossing defined by Eqs. (2.118). To get a closed system, we also need the other Y-function, which is the new Y-function Y_1^{n} introduced after the wall-crossing. Together with these new functions, the Fock–Goncharov coordinates satisfy the TBA equations (2.121)–(2.125) with $r = 2$, which are those in the maximal chamber. It is worth noting that from the pole structure of TBA equations, we are able to read off the location of the flip/discontinuity of the Stokes graph.

We then consider the Fock–Goncharov coordinates of the first and the third WKB triangulation in Fig. 2.23 and denote them by $\mathcal{X}^{(1),\text{in}}_E$ and $\mathcal{X}^{(3),\text{in}}_E$ respectively. They are related to the Y-functions by

$$\mathcal{X}^{(1),\text{in}}_{E(Q_1,Q_4)} = Y_2^{\text{n}}(e^{-\frac{\pi i}{2}}\hbar), \quad \mathcal{X}^{(1),\text{in}}_{E(Q_2,Q_4)} = Y_1(\hbar) \tag{2.173}$$

and

$$\mathcal{X}^{(3),\text{in}}_{E(Q_0,Q_2)} = Y^{\text{n}}_1(\hbar), \quad \mathcal{X}^{(3),\text{in}}_{E(Q_0,Q_3)} = Y^{\text{n}}_{12}(\hbar)^{-1}. \tag{2.174}$$

Here Y^{n} are the Y-functions after the wall-crossing in Eqs. (2.118). One can check that the discontinuity of the Fock–Goncharov coordinates coincides with the one derived from the TBA equations.

Through the Fock–Goncharov coordinates, we have seen the different sets of the Y-functions in each chamber. The marginal stability wall is located at

$$\text{Im}\left(\frac{\Pi^{(0)}_{\gamma_1}}{\Pi^{(0)}_{\gamma_2}}\right) = \left|\frac{\Pi^{(0)}_{\gamma_1}}{\Pi^{(0)}_{\gamma_2}}\right| \sin\left(\phi_1 - \phi_2 + \frac{\pi}{2}\right). \tag{2.175}$$

Therefore, the marginal statbility wall is located at $|\phi_2 - \phi_1| = \frac{\pi}{2}$, which coincides with the situation of Sect. 2.6.

2.9   Summary

In this chapter, we have reviewed the exact WKB analysis for the one-dimensional Schrödinger equation with arbitrary polynomial potential. The exact WKB period is characterized by its asymptotics and discontinuity. We have shown that the exact WKB periods satisfy the TBA equations. We then generalized the ODE/IM correspondence to the case of arbitrary polynomial potential and derived the TBA equations. In particular, the Y-functions satisfying TBA equations are shown to be identified to the Borel resummed WKB periods. The effective central charge provides strong constraints to the classical WKB periods and their quantum corrections, which is known as the PNP relation. Varying the moduli parameters of the potential, one has to modify the TBA equations, wall-crossing of the TBA equations, whose Y-functions correspond to the Borel resummed WKB periods passing the singularity. We have presented the details of the wall-crossing for the example of the quartic potential. Moreover, combined with the exact quantization condition, our TBA equations provide a powerful method to solve the spectral problem. See also [8, 22] for the case of the PT-symmetric Scrödinger equations and the higher degree polynomial.

The quantization condition of the Schrödinger equations has been studied since the foundation of quantum mechanics. In particular, the all-order Bohr–Sommerfeld quantization condition has been studied in [42–45]. The exact quantization condition for double well potential was first conjectured in [14–16] based on the instanton calculus, and then was reproduced by using the connection formula [6, 17]. See also [18] for a review. More recently, the non-perturbative expression of the energy spectrum in the quantum mechanical system has been extensively studied by using resurgence, uniform WKB, Bender–Wu, Picard–Lefschetz theory, and so on [26, 30, 46–54], where the non-perturbative structure of instanton, bion, and the PNP relations have been well studied.

The ODE/IM correspondence has been extended to the Schrödinger equation with regular singularity. In [55], the Schrödinger equation with regular singularity has been studied

$$\left(-\hbar^2\frac{d^2}{dx^2} + x^{r+1} + \sum_{a=1}^{r+2} u_a x^{r+1-a} + \hbar^2\frac{l(l+1)}{x^2}\right)\psi(x) = 0, \qquad (2.176)$$

which is a polynomial generalization of the one in Sect. 1.2. In the minimal chamber, we have found a D_{r+3}-type TBA equation. This provides a non-trivial generalization of ODE/IM correspondence and presents a solution to the Voros' Riemann–Hilbert problem for $-1 \le l \le 0$. Investigating the analytic continuation, the TBA equations have been generalized to $l > 0$ [56]. See also [57] for the correspondence of the deformed supersymmetric quantum mechanics.

The exact WKB analysis has been generalized to the higher–order ODE [9, 58–62], where a new type of Stokes line has been discovered. The Schrödinger equation (2.2) has generalized to the higher–order version [63, 64]

$$(-1)^r\epsilon^{r+1}\frac{d^{r+1}}{dx^{r+1}}\psi(x) + p(x)\psi(x) = 0, \quad p(x, \{u_i\}) = \sum_{i=0}^{N+1} u_{N+1-i}x^i, \quad (2.177)$$

which is denoted as the (A_r, A_N)-type ODE. Following a strategy similar to that in Sect. 2.4, we can derive the (A_r, A_N) Y-system. From the WKB analysis, we can calculate the asymptotic behaviors of the Y-functions, which finally give rise to the TBA equations. In a generic case, the period integral is difficult to evaluate. In the quadratic potential, (A_r, A_1)-type ODE, the TBA equations have been derived in [63], which has the form of (A_r, A_1)-type. Interestingly, the resulting TBA equations have the form as the those in (1.96), which implies the duality between (A_r, A_1) ODE and the (A_1, A_r) ODE. In [64], we have studied the WKB periods for the third-order ODE with polynomial potential, (A_2, A_N)-type ODE. In the minimal chamber of the moduli space, the Y-system and the TBA equations have the form of (A_2, A_N) type. The exact WKB periods are identified with the Y-functions of the TBA equations. Varying the moduli parameters u_i of the potential, the wall-crossing occurs similarly to Sect. 2.6. At the monomial point of the moduli space, a discrete symmetry between the WKB periods/Y-functions is enhanced, and the TBA equations obtained from the (A_2, A_2) and (A_2, A_3)-type ODE become the D_4 and E_6-type TBA equations, respectively.

The Schrödinger equation appears also in four-dimensional $\mathcal{N}=2$ supersymmetric gauge theory. One of the distinguished properties of gauge theory with $\mathcal{N} = 2$ super-symmetry is that the low-energy effective theory can be solved exactly, from which one can study strong coupling physics. The Seiberg–Witten (SW) theory [65] gives the solution of the low-energy effective action in the Coulomb branch of the moduli space in terms of the Riemann surface called the SW curve. The mass spectrum of the BPS states, which are protected against quantum corrections due to supersymmetry, are evaluated by the period integrals of the meromorphic differentials on the SW curve, called the SW differential. The Riemann surface is regarded as the spectral curve of the classical integrable model, which has the Poisson structure associated with the Seiberg–Witten differential. At some locus in the moduli space, the SW curve degenerates by shrinking the cycles, where massless monopoles and/or dyons appear. When mutually non-local massless BPS particles coexist, the theory becomes non-local. Such a theory is called the Argyres–Douglas theory [66], which gives an example of non-trivial superconformal field theories in four dimensions.

The partition function of $\mathcal{N} = 2$ gauge theory can be evaluated exactly by the localization formula for the path-integral, known as the Nekrasov partition function [67]. The precise definition of the partition function requires a regularization that the theory is defined in the omega background. In the Nekrasov–Shatashivi limit of the omega background, the classical integrable model defined from the SW differential is quantized, where the non-zero omega deformation parameter plays the role of Planck constant [68]. In this limit, the SW curve is quantized and takes the form of the Baxter T-Q relation/Schrödinger equation. The WKB period also appears as a solution of the quantum SW period of the Seiberg–Witten theory, which satisfies the TBA equations.

There is another route to obtain the differential equations associated with $\mathcal{N} = 2$ theories. The four-dimensional $\mathcal{N} = 2$ theory can be obtained by twisted compactification of the six-dimensional $\mathcal{N} = (2, 0)$ theory on a Riemann surface Σ [39]. We consider further compactification of $\mathcal{N} = 2$ theory on S^1. This provides an effective description of the three-dimensional theory, which is an $\mathcal{N} = 4$ supersymmetric sigma model whose target space is the moduli space of the solutions to the Hitchin system $(A, \varphi, \bar{\varphi})$ on Σ [69], where A and φ are the gauge connection and Higgs field respectively. See [39, 70] for a review. The characteristic polynomial of φ in the Hitchin system leads to the SW curve of the corresponding gauge theory, which thus provides a natural realization of the appearance of an integrable model in the gauge theory.

References

1. R. Balian, G. Parisi, A. Voros, Quartic oscillator, in *Mathematical Problems in Feynman Path Integral*, pp. 337–360. 5 (1978)
2. A. Voros, *Spectre de l'équation de Schrödinger et méthode BKW*. Publications Mathématiques d'Orsay (1981)
3. A. Voros, The return of the quartic oscillator. The complex WKB method. Annales de l'I.H.P. Physique théorique **39**(3), 211–338 (1983)
4. T. Aoki, T. Kawai, Y. Takei, The Bender-Wu Analysis and the Voros Theory, in *ICM-90 Satellite Conference Proceedings*. ed. by M. Kashiwara, T. Miwa (Tokyo, Springer Japan, 1991), pp.1–29
5. H. Dillinger, E. Delabaere, F. Pham, Résurgence de Voros et périodes des courbes hyperelliptiques. Annales de l'Institut Fourier **43**(1), 163–199 (1993). https://doi.org/10.5802/aif.1326
6. E. Delabaere, H. Dillinger, F. Pham, Exact semiclassical expansions for one-dimensional quantum oscillators. J. Mathem. Phys. **38**(12), 6126–6184 (1997). https://doi.org/10.1063/1.532206

7. E. Delabaere, F. Pham, Resurgent methods in semi-classical asymptotics. Annales de l'I.H.P. Physique théorique **71**(1), 1–94 (1999)
8. K. Ito, M. Mariño, H. Shu, TBA equations and resurgent Quantum Mechanics. JHEP **01**, 228 (2019). https://doi.org/10.1007/JHEP01(2019)228, arXiv:1811.04812 [hep-th]
9. N. Honda, T. Kawai, Y. Takei, *Virtual Turning Points*. Vol. 4 of SpringerBriefs in Mathematical Physics. Springer Tokyo (2015)
10. T. Kawai, Y. Takei, *Algebraic analysis of singular perturbation theory*, vol. 227. American Mathematical Soc. (2005)
11. H. Farkas, I. Kra, *Riemann Surfaces*. Graduate Texts in Mathematics. Springer, New York (1991). https://books.google.co.jp/books?id=bGZDU9HxgpAC
12. K. Iwaki, T. Nakanishi, Exact WKB analysis and cluster algebras. J. Phys. A **47**(47), 474009 (2014)
13. A.Neitzke, swn-plotter, http://www.ma.utexas.edu/users/neitzke/mathematica/swn-plotter.nb
14. J. Zinn-Justin, Multi - Instanton Contributions in Quantum Mechanics. 2. Nucl. Phys. B **218**, 333–348 (1983). https://doi.org/10.1016/0550-3213(83)90369-3
15. J. Zinn-Justin, U.D. Jentschura, Multi-instantons and exact results I: Conjectures, WKB expansions, and instanton interactions. Annals Phys. **313**, 197–267 (2004). https://doi.org/10.1016/j.aop.2004.04.004, arXiv:quant-ph/0501136 [quant-ph]
16. J. Zinn-Justin, U.D. Jentschura, Multi-instantons and exact results II: specific cases, higher-order effects, and numerical calculations. Annals Phys. **313**, 269–325 (2004). https://doi.org/10.1016/j.aop.2004.04.003, arXiv:quant-ph/0501137 [quant-ph]
17. G. Álvarez, Langer–Cherry derivation of the multi-instanton expansion for the symmetric double well. J. Math. Phys. **45**(8), 3095–3108 (2004). https://doi.org/10.1063/1.1767988
18. M. Mariño, *Advanced Topics in Quantum Mechanics*. (Cambridge University Press, 2021). https://doi.org/10.1017/9781108863384
19. Y. Sibuya, *Global theory of a second order linear ordinary differential equation with a polynomial coefficient*. North-Holland, North-Holland Mathematics Studies, Vol. 18 (1975)
20. L.F. Alday, J. Maldacena, A. Sever, P. Vieira, Y-system for scattering amplitudes.J. Phys. A **43**, 485401 (2010). https://doi.org/10.1088/1751-8113/43/48/485401, arXiv:1002.2459 [hep-th]
21. Y. Hatsuda, K. Ito, K. Sakai, Y. Satoh, Thermodynamic Bethe Ansatz equations for minimal surfaces in AdS_3. JHEP **04**, 108 (2010). https://doi.org/10.1007/JHEP04(2010)108, arXiv:1002.2941 [hep-th]
22. Y. Emery, TBA equations and quantization conditions. JHEP **07**, 171 (2021). https://doi.org/10.1007/JHEP07(2021)171, arXiv:2008.13680 [hep-th]
23. A.B. Zamolodchikov, Thermodynamic Bethe Ansatz in relativistic models. Scaling three state potts and lee-yang models. Nucl. Phys. B **342**, 695–720 (1990). https://doi.org/10.1016/0550-3213(90)90333-9
24. A.N. Kirillov, Identities for the Rogers dilogarithm function connected with simple Lie algebras. Zapiski Nauchnykh Seminarov POMI **164**, 121–133 (1987)
25. T.R. Klassen, E. Melzer, Purely elastic scattering theories and their ultraviolet limits. Nucl. Phys. B **338**, 485–528 (1990). https://doi.org/10.1016/0550-3213(90)90643-R
26. G.V. Dunne, M. Unsal, Uniform WKB, multi-instantons, and resurgent trans-series. Phys. Rev. D **89** no. 10, 105009 (2014). https://doi.org/10.1103/PhysRevD.89.105009, arXiv:1401.5202 [hep-th]
27. A. Gorsky, A. Milekhin, RG-Whitham dynamics and complex Hamiltonian systems. Nucl. Phys. B **895**, 33–63 (2015). https://doi.org/10.1016/j.nuclphysb.2015.03.028, arXiv:1408.0425 [hep-th]
28. G. Başar, G.V. Dunne, Resurgence and the Nekrasov-Shatashvili limit: connecting weak and strong coupling in the Mathieu and Lamé systems. JHEP **02**, 160 (2015). https://doi.org/10.1007/JHEP02(2015)160, arXiv:1501.05671 [hep-th]
29. S. Codesido, M. Marino, Holomorphic anomaly and quantum mechanics. J. Phys. A **51**(5), 055402 (2018). https://doi.org/10.1088/1751-8121/aa9e77, arXiv:1612.07687 [hep-th]
30. G. Basar, G.V. Dunne, M. Unsal, Quantum geometry of resurgent perturbative/nonperturbative relations. JHEP **05**, 087 (2017). https://doi.org/10.1007/JHEP05(2017)087, arXiv:1701.06572 [hep-th]

31. M. Matone, Instantons and recursion relations in N=2 SUSY gauge theory. Phys. Lett. B **357**, 342–348 (1995). https://doi.org/10.1016/0370-2693(95)00920-G, arXiv:hep-th/9506102
32. J. Sonnenschein, S. Theisen, S. Yankielowicz, On the relation between the holomorphic prepotential and the quantum moduli in SUSY gauge theories. Phys. Lett. B **367**, 145–150 (1996). https://doi.org/10.1016/0370-2693(95)01399-7, arXiv:hep-th/9510129
33. T. Eguchi , S.-K. Yang, Prepotentials of N=2 supersymmetric gauge theories and soliton equations. Mod. Phys. Lett. A **11**, 131–138 (1996). https://doi.org/10.1142/S0217732396000151, arXiv:hep-th/9510183
34. P. Dorey, C. Dunning, R. Tateo, The ODE/IM Correspondence. J. Phys. A **40**, R205 (2007). https://doi.org/10.1088/1751-8113/40/32/R01, arXiv:hep-th/0703066
35. V.V. Bazhanov, S.L. Lukyanov, A.B. Zamolodchikov, Integrable structure of conformal field theory, quantum KdV theory and thermodynamic Bethe ansatz. Commun. Math. Phys. **177**, 381–398 (1996). https://doi.org/10.1007/BF02101898, arXiv:hep-th/9412229
36. M. Marino, Nonperturbative effects and nonperturbative definitions in matrix models and topological strings. JHEP **12**, 114 (2008). https://doi.org/10.1088/1126-6708/2008/12/114, arXiv:0805.3033 [hep-th]
37. I. Aniceto, R. Schiappa, Nonperturbative Ambiguities and the reality of resurgent transseries. Commun. Math. Phys. **335**(1), 183–245 (2015). https://doi.org/10.1007/s00220-014-2165-z, arXiv:1308.1115 [hep-th]
38. R. Yaris, J. Bendler, R.A. Lovett, C.M. Bender, P.A. Fedders, Resonance calculations for arbitrary potentials. Phys. Rev. A **18**, 1816–1825 (1978). https://doi.org/10.1103/PhysRevA.18.1816
39. D. Gaiotto, G.W. Moore, A. Neitzke, Wall-crossing, Hitchin systems, and the WKB approximation. Adv. Math. **234**, 239–403 (2013). https://doi.org/10.1016/j.aim.2012.09.027, arXiv:0907.3987 [hep-th]
40. D. Gaiotto, Opers and TBA. arXiv:1403.6137 [hep-th]
41. S. Cecotti, M. Del Zotto, Y systems, Q systems, and 4D $\mathcal{N} = 2$ supersymmetric QFT. J. Phys. A **47**(47), 474001 (2014). https://doi.org/10.1088/1751-8113/47/47/474001, arXiv:1403.7613 [hep-th]
42. J.L. Dunham, The Wentzel-Brillouin-Kramers method of solving the wave equation. Phys. Rev. **41**, 713–720 (1932). https://doi.org/10.1103/PhysRev.41.713
43. C.M. Bender, K. Olaussen, P.S. Wang, Numerological analysis of the WKB approximation in large order. Phys. Rev. D **16**, 1740–1748 (1977). https://doi.org/10.1103/PhysRevD.16.1740
44. M. Robnik, L. Salasnich, WKB to all orders and the accuracy of the semiclassical quantization. J. Phys. A: Mathem. General **30**(5), 1711 (1997). https://doi.org/10.1088/0305-4470/30/5/031
45. M. Robnik, V.G. Romanovski, Some properties of WKB series. J. Phys. A: Mathem. General **33**(28), 5093 (2000). https://doi.org/10.1088/0305-4470/33/28/312
46. G.V. Dunne, M. Ünsal, Generating nonperturbative physics from perturbation theory. Phys. Rev. D **89**(4), 041701 (2014). https://doi.org/10.1103/PhysRevD.89.041701, arXiv:1306.4405 [hep-th]
47. Y. Tanizaki, T. Koike, Real-time Feynman path integral with Picard–Lefschetz theory and its applications to quantum tunneling. Annals Phys. **351**, 250–274 (2014). https://doi.org/10.1016/j.aop.2014.09.003, arXiv:1406.2386 [math-ph]
48. A. Cherman, M. Unsal, Real-time Feynman path integral realization of instantons. arXiv:1408.0012 [hep-th]
49. A. Behtash, G.V. Dunne, T. Schaefer, T. Sulejmanpasic, M. Ünsal, Critical points at infinity, non-gaussian saddles, and Bions. JHEP **06**, 068 (2018). https://doi.org/10.1007/JHEP06(2018)068, arXiv:1803.11533 [hep-th]
50. N. Sueishi, $1/\epsilon$ problem in resurgence. PTEP **2021**(1), 013B01 (2021). https://doi.org/10.1093/ptep/ptaa156, arXiv:1912.03518 [hep-th]
51. G.V. Dunne, T. Sulejmanpasic, M. Ünsal, Bions and instantons in triple-well and multi-well potentials. arXiv:2001.10128 [hep-th]
52. N. Sueishi, S. Kamata, T. Misumi, M. Ünsal, On exact-WKB analysis, resurgent structure, and quantization conditions. JHEP **12**, 114 (2020). https://doi.org/10.1007/JHEP12(2020)114, arXiv:2008.00379 [hep-th]

53. N. Sueishi, S. Kamata, T. Misumi, M. Ünsal, Exact-WKB, complete resurgent structure, and mixed anomaly in quantum mechanics on S^1. JHEP **07**, 096 (2021). https://doi.org/10.1007/JHEP07(2021)096, arXiv:2103.06586 [quant-ph]

54. S. Kamata, T. Misumi, N. Sueishi, M. Ünsal, Exact WKB analysis for SUSY and quantum deformed potentials: quantum mechanics with Grassmann fields and Wess-Zumino terms. Phys. Rev. D **107**(4), 045019 (2023). https://doi.org/10.1103/PhysRevD.107.045019, arXiv:2111.05922 [hep-th]

55. K. Ito, H. Shu, TBA equations for the Schrödinger equation with a regular singularity. J. Phys. A **53**(33), 335201 (2020). https://doi.org/10.1088/1751-8121/ab96ee, arXiv:1910.09406 [hep-th]

56. B. Gabai, X. Yin, Exact quantization and analytic continuation. arXiv:2109.07516 [hep-th]

57. K. Ito, H. Shu, TBA equations and exact WKB analysis in deformed supersymmetric quantum mechanics. JHEP **03**, 122 (2024). https://doi.org/10.1007/JHEP03(2024)122, arXiv:2401.03766 [hep-th]

58. H.L. Berk, W.M. Nevins, K.V. Roberts, New Stokes" line in WKB theory. J. Mathem. Phys. **23**(6), 988–1002 (1982). https://doi.org/10.1063/1.525467

59. T.K. T. Aoki, Y. Takei, New turning points in the exact WKB analysis for higher- order ordinary differential equations. Analyse algebrique des perturbations singulieres, I, Methodes resurgentes, 69–84 (1994)

60. T. Aoki, T. Kawai, Y. Takei, On the exact WKB analysis for the third order ordinary differential equations with a large parameter. Asian J. Mathem. **2**(4), 625–640 (1998)

61. L. Hollands, A. Neitzke, Exact WKB and abelianization for the T_3 equation. Commun. Math. Phys. **380**(1), 131–186 (2020). https://doi.org/10.1007/s00220-020-03875-1, arXiv:1906.04271 [hep-th]

62. F. Yan, Exact WKB and the quantum Seiberg-Witten curve for 4d N = 2 pure SU(3) Yang-Mills. Abelianization. JHEP **03**, 164 (2022). https://doi.org/10.1007/JHEP03(2022)164, arXiv:2012.15658 [hep-th]

63. K. Ito, T. Kondo, K. Kuroda, H. Shu, WKB periods for higher order ODE and TBA equations. JHEP **10**, 167 (2021). https://doi.org/10.1007/JHEP10(2021)167, arXiv:2104.13680 [hep-th]

64. K. Ito, T. Kondo, H. Shu, Wall-crossing of TBA equations and WKB periods for the third order ODE. Nucl. Phys. B **979**, 115788 (2022). https://doi.org/10.1016/j.nuclphysb.2022.115788, arXiv:2111.11047 [hep-th]

65. N. Seiberg, E. Witten, Electric – magnetic duality, monopole condensation, and confinement in N=2 supersymmetric Yang-Mills theory. Nucl. Phys. B **426**, 19–52 (1994). https://doi.org/10.1016/0550-3213(94)90124-4, arXiv:hep-th/9407087. [Erratum: Nucl. Phys. B 430, 485–486 (1994)]

66. P.C. Argyres, M.R. Douglas, New phenomena in SU(3) supersymmetric gauge theory. Nucl. Phys. **B448**, 93–126 (1995). https://doi.org/10.1016/0550-3213(95)00281-V, arXiv:hep-th/9505062 [hep-th]

67. N.A. Nekrasov, Seiberg-Witten prepotential from instanton counting. Adv. Theor. Math. Phys. **7**(5), 831–864 (2003). https://doi.org/10.4310/ATMP.2003.v7.n5.a4, arXiv:hep-th/0206161 [hep-th]

68. N.A. Nekrasov, S.L. Shatashvili, Quantization of Integrable Systems and Four Dimensional Gauge Theories, in *16th International Congress on Mathematical Physics*, pp. 265–289. 8, (2009). https://doi.org/10.1142/9789814304634_0015. arXiv:0908.4052 [hep-th]

69. N.J. Hitchin, The self-duality equations on a Riemann surface. Proc. London Mathem. Soc. **s3-55**(1), 59–126 (1987). https://doi.org/10.1112/plms/s3-55.1.59

70. A. Neitzke, *Hitchin Systems in N = 2 Field Theory*, pp. 53–77 (2016). https://doi.org/10.1007/978-3-319-18769-3-3, arXiv:1412.7120 [hep-th]

Chapter 3
Massive ODE/IM Correspondence

In Chap. 2, we have seen the correspondence between ordinary differential equations and two-dimensional conformal field theories (CFTs). CFT is regarded as a massless limit of a massive integrable model. A massive version of the correspondence between the differential equations and massive integrable models has been proposed by Lukyanov and Zamolodchikov [1], who studied the linear problem for the modified sinh-Gordon equation and its relation to the quantum sine-Gordon model. They defined the Q-functions as the connections coefficients of the solutions around the infinity and the origin of the complex plane, which are shown to satisfy the same quantum Wronskian relations and the asymptotics as well as analyticity for the spectral parameter, of the quantum sine-Gordon models. This correspondence relates the classical integrals of motions in the Sinh-Gordon model defined by the Lax operator formalism to the quantum integrals of motions in the quantum sine-Gordon model, whose values are evaluated on a certain sector of the Hilbert space.

The sinh-Gordon equation is a fundamental integrable equation that appears in various contexts. It belongs to a class of the affine Toda field equation associated with the affine Lie algebra $A_1^{(1)}$ [2]. This massive ODE/IM correspondence has been generalized to affine Toda field equations associated with an affine Lie algebra $\hat{\mathfrak{g}}$ in [3–6]. The related integrable models are found to be the integrable models with the BAEs associated with the Langlands dual $\hat{\mathfrak{g}}^\vee$ of $\hat{\mathfrak{g}}$.

The linear problem associated with the modified affine Toda field equation is made of two differential equations, holomorphic and anti-holomorphic. In the light-cone limit, where the anti-holomorphic part is eliminated, one obtains a linear differential system that reduces to a higher-order ordinary differential equation whose ODE/IM correspondence was studied in [7]. The ODE/IM correspondence based on the linear system has been studied in [8–11].

The Sinh-Gordon equation appears also as the equation for the minimal surface in the three-dimensional ant-de Sitter spacetime AdS_3 [12]. It is generalized to higher dimensional AdS spacetime, where the minimal surface for AdS_4 is related to $B_2^{(1)}$

© The Author(s), under exclusive licence to Springer Nature Singapore Pte Ltd. 2025
K. Ito and H. Shu, *ODE/IM Correspondence and Quantum Periods*, SpringerBriefs in Mathematical Physics 51, https://doi.org/10.1007/978-981-96-0499-9_3

affine Toda field equations [13]. The area of the minimal surface with a boundary is one of the important quantities to test the AdS/CFT correspondence. In particular, the minimal surface whose boundary is composed of light-like segments in AdS space-time corresponds to the gluon-scattering amplitude whose momenta are proportional to the segments or the Wilson loop with the same boundary [14]. The area of the minimal surface is the strong coupling limit of the gluon scattering amplitude. A remarkable observation by Alday–Gaiotto–Maldacena is that the area of the surface is determined by the TBA-like equations [15]. The T- and Y-functions are determined by the Wronskian of the subdominant solutions of the linear problem. For AdS_3 the TBA equations have been formulated by [16], which is the same form as the homogeneous sine-Gordon model [17]. The TBA equations also give the massive version of the ODE/IM correspondence with polynomial potential. The massive version of the TBA equations appears also in the WKB analysis of the Hitchin system in $\mathcal{N} = 2$ supersymmetric gauge theories [18], where the Y-functions are identified with the Fock–Goncharov coordinates of the moduli space of the Hitchin system.

In this chapter, we review the massive ODE/IM correspondence and its relation to the massless ODE/IM. In Sect. 3.1, the modified affine Toda field equations, and their associated linear problems are presented. We construct the asymptotic solutions of the linear problem and define the Q-functions as the connection coefficients. The ψ-system, which connects the solutions in different representations, is formulated to write down the Bethe ansatz equations. In Sect. 3.2, Baxter's T-Q relations and the NLIE are formulated. In Sect. 3.3, the light-cone limit and the relation to the ODE/IM correspondence are discussed.

3.1 Modified Affine Toda Field Equations

3.1.1 Affine Lie Algebra

We start by explaining some basic definitions of Lie algebra. Let $\mathfrak{g}$ be a simple Lie algebra of rank r. Denote $\{E_\alpha, H^i\}$ ($\alpha \in \Delta$, $i = 1, \ldots r$) the generators of $\mathfrak{g}$, where $\Delta = \Delta_+ \cup \Delta_-$ is the set of roots and Δ_+ (Δ_-) is the set of positive (negative) roots. The generators satisfy the commutation relations

$$[E_\alpha, E_\beta] = \begin{cases} N_{\alpha,\beta} E_{\alpha+\beta}, & \text{for } \alpha + \beta \in \Delta, \\ \alpha^\vee \cdot H & \alpha + \beta = 0, \\ 0, & \text{otherwise.} \end{cases} \tag{3.1}$$

$$[H^i, E_\alpha] = \alpha^i E_\alpha, \tag{3.2}$$

$$[H^i, H^j] = 0, \tag{3.3}$$

where we define the coroot of α by $\alpha^\vee = 2\alpha/\alpha^2$. Let $\alpha_1, \ldots, \alpha_r$ be the simple roots of $\mathfrak{g}$. The Cartan matrix of $\mathfrak{g}$ is defined by $A_{ij} = \alpha_i \cdot \alpha_j^\vee$. We normalize the long root whose squared length is 2. The fundamental (co)weight vectors ω_i ($\omega_i^\vee$) ($i = 1, \ldots, r$) are defined to be dual to the simple roots: $\omega_i \cdot \alpha_j^\vee = \delta_{ij}$ ($\omega_i^\vee \cdot \alpha_j = \delta_{ij}$). Denote the (co)highest root by θ ($\theta^\vee$), which is expanded as

$$\theta = \sum_{i=1}^r n_i \alpha_i, \quad \theta^\vee = \sum_{i=1}^r n_i^\vee \alpha_i^\vee \tag{3.4}$$

with non-negative integers n_i ($n_i^\vee$) called the (dual) Coxeter labels. The algebraic structure of the Lie algebra is determined by the Dynkin diagram. We define the (dual) Coxeter number h ($h^\vee$) by

$$h = 1 + \sum_{i=1}^r n_i, \quad h^\vee = 1 + \sum_{i=1}^r n_i^\vee. \tag{3.5}$$

We also define the (co-)Weyl vector by

$$\rho = \sum_{i=1}^r \omega_i, \quad \rho^\vee = \sum_{i=1}^r \omega_i^\vee. \tag{3.6}$$

The affine Lie algebra $\hat{\mathfrak{g}}$ of a simple Lie algebra $\mathfrak{g}$ is a central extension by an element k called the level, of $\mathfrak{g} \otimes \mathbb{C}[t, t^{-1}]$: $\hat{\mathfrak{g}} = \mathfrak{g} \otimes \mathbb{C}[t, t^{-1}] \oplus \mathbb{C}k$, where the commutators of two elements $a \otimes t^n, b \otimes t^m \in \hat{\mathfrak{g}}$ ($a, b \in \mathfrak{g}, m, n \in \mathbb{Z}$) is defined by

$$[a \otimes t^n, b \otimes t^m] = [a, b] \otimes t^{m+n} + (a, b)n\delta_{m+n,0}k. \tag{3.7}$$

Here (a, b) denotes the Killing form on $\mathfrak{g}$. The Dynkin diagram of $\hat{\mathfrak{g}}$ is obtained by adding the root $\alpha_0 = -\theta$ to the that of $\mathfrak{g}$. Defining the Coxeter label for α_0 by $n_0 = 1$, we have $\sum_{i=0}^r n_i \alpha_i = 0$. The above algebra $\hat{\mathfrak{g}}$ is also called the untwisted affine Lie algebra denoted by $\mathfrak{g}^{(1)}$. When $\mathfrak{g}$ admits an outer automorphism, one can define the twisted affine Lie algebra $\mathfrak{g}^{(r)}$ which has a different extended simple root.

We denote the Langlands dual $\hat{\mathfrak{g}}^\vee$ of an affine Lie algebra $\hat{\mathfrak{g}}$, whose simple roots are given by $\alpha_i^\vee$ ($i = 0, \ldots, r$). For the simply-laced Lie algebra $\mathfrak{g} = A_r, D_r, E_6, E_7, E_8$, the simple roots have squared lengths of 2. Then $A_r^{(1)}, D_r^{(1)}, E_r^{(1)}$ are self-dual, whereas for a non-simply-laced $\mathfrak{g}^{(1)}$ they are not self-dual: $(B_r^{(1)})^\vee = A_{2r-1}^{(2)}, (C_r^{(1)})^\vee = D_{r+1}^{(2)}$, $(F_4^{(1)})^\vee = E_6^{(2)}$ and $(G_2^{(1)})^\vee = D_4^{(3)}$.

3.1.2 Modified Affine Toda Field Equation

We discuss the **affine Toda field theory** associated with an affine Lie algebra $\hat{\mathfrak{g}}$. The theory is defined on two-dimensional Euclidean spacetime with coordinates $x^\mu = (x^0, x^1)$. We also use complex coordinates $(z, \bar{z})$ defined by

$$z = \frac{1}{2}(x^0 + ix^1), \quad \bar{z} = \frac{1}{2}(x^0 - ix^1). \tag{3.8}$$

Note that $\partial = \partial_z = \partial_0 - i\partial_1$ and $\bar{\partial} = \partial_{\bar{z}} = \partial_0 + i\partial_1$. The Lagrangian of the affine Toda field theory based on $\hat{\mathfrak{g}}$ is given by

$$\mathcal{L} = \frac{1}{2}\partial^\mu\phi \cdot \partial_\mu\phi + \left(\frac{m}{\beta}\right)^2 \sum_{i=0}^{r} n_i \left[\exp(\beta\alpha_i\phi) - 1\right], \tag{3.9}$$

where $\phi = (\phi_1, \dots, \phi_r)$ are r-component scalar fields, m, and β are mass and coupling parameters. α_i are simple roots of $\hat{\mathfrak{g}}$ and n_i are the Coxeter labels. The equation of motion is

$$\partial^\mu\partial_\mu\phi - \left(\frac{m^2}{\beta}\right) \sum_{i=0}^{r} n_i\alpha_i \exp(\beta\alpha_i\phi) = 0. \tag{3.10}$$

We call this equation the affine Toda field equation.

The affine Toda field Eq. (3.10) is written in the form of the zero curvature condition for the connection A and $\bar{A}$

$$[\partial + A, \bar{\partial} + \bar{A}] = 0, \tag{3.11}$$

where

$$A = \frac{\beta}{2}\partial\phi \cdot H - me^\lambda \sum_{i=0}^{r} \sqrt{n_i^\vee} E_{\alpha_i} \exp\left(\frac{\beta}{2}\alpha_i \cdot \phi\right), \tag{3.12}$$

$$\bar{A} = -\frac{\beta}{2}\bar{\partial}\phi \cdot H - me^{-\lambda} \sum_{i=0}^{r} \sqrt{n_i^\vee} E_{-\alpha_i} \exp\left(\frac{\beta}{2}\alpha_i \cdot \phi\right). \tag{3.13}$$

Here we introduced the spectral parameter λ. The zero curvature condition (3.11) is equivalent to the integrability condition of the linear problem:

$$(\partial + A)\psi = 0, \quad (\bar{\partial} + \bar{A})\psi = 0. \tag{3.14}$$

Here ψ is a vector that belongs to a representation of $\mathfrak{g}$.

The non-affine version of the Toda field equation based on a Lie algebra $\mathfrak{g}$ is obtained by removing the term $e^{\beta\alpha_0\phi}$ from the Lagrangian (3.9). Its action and the field equation are invariant under the conformal transformation

$$z \to \tilde{z} = f(z), \quad \phi(z, \bar{z}) \to \tilde{\phi}(\tilde{z}, \bar{\tilde{z}}) = \phi(z, \bar{z}) - \frac{1}{\beta}\rho^\vee \log(\partial f \bar{\partial}\bar{f}), \tag{3.15}$$

where $f(z)$ is an analytic function of z and $\bar{f}(\bar{z})$ is its complex conjugate. The affine Toda field Eq. (3.10), however, is not invariant under the conformal transformation.

It transforms to

$$\partial \bar{\partial} \phi - \left(\frac{m^2}{\beta} \right) \left[\sum_{i=1}^{r} n_i \alpha_i \exp(\beta \alpha_i \phi) + p(z) \bar{p}(\bar{z}) \alpha_0 \exp(\beta \alpha_0 \phi) \right] = 0, \qquad (3.16)$$

where we have defined

$$p(z) = (\partial f)^h, \quad \bar{p}(\bar{z}) = (\bar{\partial} \bar{f})^h. \qquad (3.17)$$

We refer to the Eq. (3.16) as the **modified affine Toda field equation**. This can be written in the form of the zero curvature condition $[\mathcal{D}, \bar{\mathcal{D}}] = 0$ where

$$\mathcal{D} = \partial + \mathcal{A}, \quad \bar{\mathcal{D}} = \bar{\partial} + \bar{\mathcal{A}} \qquad (3.18)$$

and

$$\begin{aligned}
\mathcal{A} &= \frac{\beta}{2} \partial \phi \cdot H - m e^{\lambda} \left\{ \sum_{i=1}^{r} \sqrt{n_i^{\vee}} E_{\alpha_i} \exp \left(\frac{\beta}{2} \alpha_i \phi \right) + (-1)^h p(z) E_{\alpha_0} \exp \left(\frac{\beta}{2} \alpha_0 \phi \right) \right\}, \\
\bar{\mathcal{A}} &= -\frac{\beta}{2} \bar{\partial} \phi \cdot H - m e^{-\lambda} \left\{ \sum_{i=1}^{r} \sqrt{n_i^{\vee}} E_{-\alpha_i} \exp \left(\frac{\beta}{2} \alpha_i \phi \right) + (-1)^h \bar{p}(\bar{z}) E_{-\alpha_0} \exp \left(\frac{\beta}{2} \alpha_0 \phi \right) \right\}.
\end{aligned}$$

$$(3.19)$$

The linear equations associated with Eq. (3.16) are given by

$$\mathcal{D} \psi = 0, \quad \bar{\mathcal{D}} \psi = 0. \qquad (3.20)$$

We will consider the case that $p(z)$ is a polynomial in z, in particular, we choose $p(z)$ as a monomial of z:

$$p(z) = z^{hM} - s^{hM}, \qquad (3.21)$$

for a positive number $M > \frac{1}{h-1}$. Here h is the Coxeter number of g.

A special family of solutions to Eq. (3.16) with the following properties are important. Let us introduce the polar coordinates (ρ, θ) by $z = \rho e^{i\theta}$.

- $p(z)$ in (3.21) has a periodicity under $\theta \to \theta + \frac{2\pi}{hM}$. We assume that the solution $\phi(\rho, \theta)$ of (3.16) satisfies the same periodicity condition:

$$\phi \left(\rho, \theta + \frac{2\pi}{hM} \right) = \phi(\rho, \theta). \qquad (3.22)$$

- For large ρ, where $p(z) \sim \rho^{hM}$, Eq. (3.16) can be solved as

$$\phi(\rho, \theta) = \frac{2M\rho^\vee}{\beta} \log \rho + O(1), \quad \rho \to \infty. \tag{3.23}$$

Here $\rho^\vee$ is the co-Weyl vector.

- For small ρ, we impose the boundary condition:

$$\phi(\rho, \theta) = 2g \log \rho + O(1), \quad \rho \to 0, \tag{3.24}$$

for some r-dimensional vector g such that $\beta\alpha_a \cdot g + 1 > 0$ $(a = 0, 1, \ldots, r)$. These conditions are required so that the log term in (3.24) gives the leading order term of the solution.

Due to the periodic property of $\phi(z, \bar{z})$ and $p(z)$, we can show that the linear problem is invariant under the rotation $\hat{\Omega}_k$ $(k \in \mathbb{Z})$:

$$\hat{\Omega}_k : \begin{cases} z & \to z e^{\frac{2\pi k i}{hM}} \\ s & \to s e^{\frac{2\pi k i}{hM}} \\ \lambda & \to \lambda - \frac{2\pi k i}{hM} \end{cases} . \tag{3.25}$$

Hence, if ψ is a solution of the linear problem, then

$$\psi_{[k]}(z, \bar{z}) := \hat{\Omega}_k \psi(z, \bar{z}) \tag{3.26}$$

is also a solution to the linear problem (3.20).

There is also the discrete $\mathbb{Z}_h$ symmetry transformation defined by

$$\hat{\Pi} : \begin{cases} \lambda & \to \lambda - \frac{2\pi}{h} \\ \mathcal{A} & \to S\mathcal{A}S^{-1} \\ \psi & \to S\psi \end{cases} , \tag{3.27}$$

where

$$S := \exp\left(\frac{2\pi i}{h} \rho^\vee \cdot H\right). \tag{3.28}$$

This symmetry follows from

$$S E_{\alpha_i} S^{-1} = e^{2\pi i/h} E_{\alpha_i}, \quad i = 0, 1, \ldots, r. \tag{3.29}$$

We now consider the asymptotic solutions of linear problem $\mathcal{D}\psi = 0$ and $\bar{\mathcal{D}}\psi = 0$ in a representation V of $\mathfrak{g}$, where ψ takes value in the vector space V. To construct the solution, it is convenient to do the gauge transformation

$$\mathcal{A}^U = U\partial U^{-1} + UAU^{-1}, \quad \bar{\mathcal{A}}^U = U\bar{\partial}U^{-1} + U\bar{A}U^{-1}, \quad \psi \to U\psi, \tag{3.30}$$

and simplify the linear problem. For example, if we choose

$$U = \exp(-\beta\phi \cdot H/2 + \frac{1}{h}\log p(z)\rho^\vee \cdot H), \tag{3.31}$$

we obtain

$$\mathcal{A}^U = \beta\partial\phi \cdot H - \frac{1}{h}\partial\log p\rho^\vee \cdot H + me^\lambda(p(z))^{\frac{1}{h}}\Lambda_+, \tag{3.32}$$

$$\bar{\mathcal{A}}^U = me^{-\lambda}p(z)^{-\frac{1}{h}}\left\{\sum_{i=1}^r \sqrt{n_i^\vee}e^{\beta\alpha_i\cdot\phi}E_{-\alpha_i} + \sqrt{n_0^\vee}p(z)\bar{p}(\bar{z})e^{\beta\alpha_0\cdot\phi}E_{-\alpha_0}\right\}. \tag{3.33}$$

Similarly, when we perform the gauge transformation with

$$U' = \exp(\beta\phi \cdot H/2 - \frac{1}{h}\log \bar{p}(\bar{z})\rho^\vee \cdot H) \tag{3.34}$$

we get

$$\mathcal{A}^{U'} = me^\lambda \bar{p}(\bar{z})^{-\frac{1}{h}}\left\{\sum_{i-1}^r \sqrt{n_i^\vee}e^{\beta\alpha_i\cdot\phi}E_{\alpha_i} + \sqrt{n_0^\vee}p(z)\bar{p}(\bar{z})e^{\beta\alpha_0\cdot\phi}E_{\alpha_0}\right\},$$

$$\bar{\mathcal{A}}^{U'} = -\beta\bar{\partial}\phi \cdot H + \frac{1}{h}\bar{\partial}\log \bar{p}\rho^\vee \cdot H + me^{-\lambda}(\bar{p}(\bar{z}))^{\frac{1}{h}}\Lambda_-. \tag{3.35}$$

Here we defined

$$\Lambda_\pm = \sum_{i=1}^r \sqrt{n_i^\vee}E_{\pm\alpha_i} + \sqrt{n_0^\vee}E_{\pm\alpha_0}. \tag{3.36}$$

For large z, we have $p(z) \sim z^{hM}$ and $\phi(z,\bar{z}) \sim \frac{M}{\beta}\rho^\vee \log(z\bar{z})$. Then the connections become

$$\mathcal{A}^U \sim me^\lambda z^M \Lambda_+, \quad \bar{\mathcal{A}}^U \sim me^{-\lambda}\left(\frac{\bar{z}}{z}\right)^M \Lambda_-. \tag{3.37}$$

We find the z-dependent part of the asymptotic solution by solving $(\partial + \mathcal{A}^U)\psi = 0$. We can also calculate the $\bar{z}$-dependent part by solving $(\bar{\partial} + \bar{\mathcal{A}}^{U'})\psi = 0$ for large $\bar{z}$. Perform the inverse gauge transformations on ψ and combining both contributions, one obtains the asymptotic solutions for large ρ.

Let us consider the linear problem on the fundamental representation $V^{(a)}$ with the highest weight ω_a. The vector space $V^{(a)}$ can be decomposed into the direct sum of the eigenspaces spanned by the eigenvector of $\Lambda_\pm$. We can easily solve the linear problem asymptotically on the eigenspace. The subdominant solution corresponds to the eigenvalue with the largest real part. We denote the eigenvalue $\mu_\pm^{(a)}$ and the eigenvector $\boldsymbol{\mu}_{\pm}^{(a)}$ of $\Lambda_{\pm}$. Let us choose the basis of the representation such that $\Lambda_- = {}^t(\Lambda_+)$ holds where the eigenvectors and the eigenvalues are the same: $\mu^{(a)} := \mu_-^{(a)} = \mu_+^{(a)}$. Then the asymptotic form of the subdominant solution is found to be for large ρ,

$$
\begin{aligned}
\Psi^{(a)}(z, \bar{z}) &= \exp\left(-\mu^{(a)} m e^{\lambda} \frac{z^{M+1}}{M+1} - \mu^{(a)} m e^{-\lambda} \frac{\bar{z}^{M+1}}{M+1} - \frac{M}{2} \rho^{\vee} \cdot H(\log z - \log \bar{z})\right) \boldsymbol{\mu}^{(a)} \\
&= \exp\left(-2\mu^{(a)} \frac{\rho^{M+1}}{M+1} m \cosh(\lambda + i\theta(M+1))\right) e^{-i\theta M \rho^{\vee} \cdot H} \boldsymbol{\mu}^{(a)}.
\end{aligned}
\tag{3.38}
$$

Applying the rotation $\hat{\Omega}_k$ to the above $\Psi^{(a)}$, we have

$$
\Psi_{[k]}^{(a)}(z, \bar{z}) = \exp\left(-2\mu^{(a)} \frac{\rho^{M+1}}{M+1} m \cosh\left(\lambda + i\theta(M+1) + \frac{2\pi i k}{h}\right)\right) e^{-i(\theta M + \frac{2\pi k}{h})\rho^{\vee} \cdot H} \boldsymbol{\mu}^{(a)},
\tag{3.39}
$$

which is subdominant in the sector

$$
S_k = \left\{\theta; \left|\theta - \frac{2\pi k}{h(M+1)}\right| < \frac{\pi}{h(M+1)}\right\}.
\tag{3.40}
$$

Now we consider the solution around $z = 0$. For small z, the relevant equation is $(\partial + \mathcal{A}^U)\tilde{\psi} = 0$, where

$$
\mathcal{A}^U = \beta g \cdot H \frac{1}{z} + \text{regular}.
\tag{3.41}
$$

Let $\mathbf{e}_i^{(a)}$ be a vector with the weight $h_i^{(a)}$ of the representation $V^{(a)}$ with $h_1^{(a)} = \omega_a$. The solution is

$$
\tilde{\psi}_i = \exp(-\beta g \cdot h_i^{(a)} \log z)\mathbf{e}_i^{(a)} + \dots.
\tag{3.42}
$$

By the inverse gauge transformation and including the anti-holomorphic part, we obtain

$$
\mathcal{X}_i^{(a)}(z, \bar{z}) = \exp\left(-\frac{\beta}{2} g \cdot h_i^{(a)}(\log z - \log \bar{z} + 2\lambda)\right) \mathbf{e}_i^{(a)} + \dots.
\tag{3.43}
$$

Here we introduce the λ dependence such that the basis function $\mathcal{X}_i^{(a)}$ is invariant under $\hat{\Omega}_k$. Since $\mathcal{X}_i^{(a)}$ forms a basis of solutions to the linear problem, the subdominant solution $\Psi^{(a)}$ can be expanded as

$$
\Psi^{(a)}(z, \bar{z}|\lambda, g) = \sum_{i=1}^{\dim(V^{(a)})} Q_i^{(a)}(\lambda, g) \, \mathcal{X}_i^{(a)}(z, \bar{z}|\lambda, g).
\tag{3.44}
$$

These coefficients $Q_i^{(a)}(\lambda, g)$ are regarded as the Q-functions of the massive quantum integrable model corresponding to the linear problem (3.20). From the above observation that $\hat{\Omega}_1 \hat{\Pi} \Psi^{(a)} = \Psi^{(a)}$ and the asymptotic form of $X_i^{(a)}$, one can derive a quasi-periodic condition for these Q-functions,

$$Q_i^{(a)} \left(\lambda - \tfrac{2\pi i}{hM}(M+1), g\right) = \exp\left(-\tfrac{2\pi i}{h}(\rho^\vee + \beta g) \cdot h_i^{(a)}\right) Q_i^{(a)}(\lambda, g). \qquad (3.45)$$

3.1.3 ψ-System and BAE

We have constructed the subdominant solution $\Psi^{(a)}$ and the basis of the solution around the origin for the representation $V^{(a)}$ $(a = 1, \ldots, r)$. The connection coefficients define the set of Q-functions. We can construct the solution for any representation of $\mathfrak{g}$ by the tensor product of the solutions on $V^{(a)}$. Since the tensor product can be decomposed into irreducible representations, one finds some non-trivial relations for the anti-symmetric product of the tensor product of two $V^{(a)}$. For A_r, we find the embedding map [19]

$$V^{(a)} \wedge V^{(a)} \hookrightarrow V^{(a-1)} \otimes V^{(a+1)}, \qquad (3.46)$$

since the highest weight vectors are the same: $2\omega_a - \alpha_a = \omega_{a-1} + \omega_{a+1}$.

In general, for the representation $V^{(a)}$ of a simply-laced algebra $\mathfrak{g}$, one has the inclusion map:

$$\iota\left(V^{(a)} \wedge V^{(a)}\right) \hookrightarrow \bigotimes_{b=1}^{r} \left(V^{(b)}\right)^{2\delta_{ab} - A_{ab}}, \quad a = 1, \ldots, r, \qquad (3.47)$$

where $V^{(0)} = V^{(r+1)} = \mathbb{C}$. We can then write the linear problem for the representations on both sides. The subdominant solutions on both sides can be expressed by using the wedge product and tensor product of the solution (3.39) respectively. To match the asymptotics of the subdominant solutions, we should rotate the solution by $\hat{\Omega}_k$ with an integer or a half-integer k. From the uniqueness of the solution to the linear problem, one finds the relation of the subdominant solutions around the positive real axis:

$$\iota\left(\Psi_{[-\frac{1}{2}]}^{(a)} \wedge \Psi_{[\frac{1}{2}]}^{(a)}\right) = \bigotimes_{b=1}^{r} \left(\Psi^{(b)}\right)^{2\delta_{ab} - A_{ab}}, \quad a = 1, \ldots, r, \qquad (3.48)$$

with $\Psi^{(0)} = \Psi^{(r+1)} = 1$, which is the ψ-system for algebra $\mathfrak{g}$ [6, 8, 9]. Here the subscript $[\pm\frac{1}{2}]$ denotes the rotation

$$f_{[\pm\frac{1}{2}]}(z,\lambda) = \hat{\Omega}_k f(z,\lambda), \quad k = \pm\frac{1}{2}. \tag{3.49}$$

Substituting the expansion of the solution (3.44) into the ψ-system, the coefficient of the highest weight state leads to

$$Q^{(a)}_{1,[-\frac{1}{2}]} \wedge Q^{(a)}_{2,[\frac{1}{2}]} - Q^{(a)}_{1,[\frac{1}{2}]} \wedge Q^{(a)}_{2,[-\frac{1}{2}]} = \prod_{b=1}^{r} \left(Q^b_1\right)^{2\delta_{ab}-A_{ab}}, \quad a = 1,\dots,r. \tag{3.50}$$

Taking the zeros $\lambda^{(a)}_j$ of $Q^{(a)}_1$, i.e. $Q^{(a)}_1(\lambda^{(a)}_j) = 0$, one obtains the Bethe ansatz equations

$$\prod_{b=1}^{r} \left.\frac{Q^{(b)}_{[A_{ab}/2]}}{Q^{(b)}_{[-A_{ab}/2]}}\right|_{\lambda^{(a)}_j} = -1, \quad a = 1,\dots,r. \tag{3.51}$$

These BAEs are the massive version of the those derived in Chap. 1.

3.2   T-Q Relation and T-/Y-System for A_r-type

In Chap. 1 we have introduced the Baxter's T-Q relations from the Wronskians of the solutions of the ODE and derived the Bethe ansatz equation from these relations. For the linear problem associated with the modified affine Toda field equation, one can define the T-Q relation from the Wronskians. Furthermore, the Wronskians of the subdominant solutions satisfy the Plücker relations which lead to the T-system and construct the Y-system. The difference from the ODE/IM correspondence arises from the asymptotic boundary conditions. In this section, we will focus on the $A^{(1)}_r$-type affine Lie algebra. We first derive Baxter's T-Q relation from the linear problem and reproduce the A_r-type Bethe ansatz equations [20]. We then introduce more generic T-functions and the counterpart Y-functions and derive a closed system satisfied by them.

3.2.1   Baxter's T-Q Relation for $A^{(1)}_r$

Let us consider the linear problem (3.20) associated with the modified affine Toda field equation for $A^{(1)}_r$ in the fundamental representation $V^{(1)}$ with the highest weight ω_1 of dimension $r + 1$. We denote $\Psi^{(1)}_k = \hat{\Omega}_{-k}\Psi^{(1)}$ the subdominant solution (3.39) in the Stokes sector $\mathcal{S}_k$ ($k \in \mathbb{Z}$) (3.40).[1] $\{\Psi^{(1)}_1, \dots \Psi^{(1)}_{r+1}\}$ spans a basis of the solutions for $V^{(1)}$. Similarly, one can construct the basis $\{\mathcal{X}^{(1)}_1, \dots, \mathcal{X}^{(1)}_{r+1}\}$ around the origin. The p-th anti-symmetric tensor product $V^{(1)} \wedge \cdots \wedge V^{(1)}$ is equivalent to $V^{(p)}$ which

---

[1] In [20], Ψ_k is denoted by s_k.

has the highest weight ω_p for $p = 2, \ldots, r$. The basis of the solutions in $V^{(p)}$ is $\{\Psi_{i_1}^{(1)} \wedge \cdots \wedge \Psi_{i_p}^{(1)}\}$, where we can admit the rotation by $\hat{\Omega}_k$ with a (half-)integer k. The subdominant solution around the positive real axis for $V^{(p)}$ is provided by

$$\Psi^{(p)} = \Psi_{[-\frac{p-1}{2}]}^{(1)} \wedge \Psi_{[-\frac{p+1}{2}]}^{(1)} \wedge \cdots \Psi_{[\frac{p-1}{2}]}^{(1)}, \tag{3.52}$$

where we have a fractional shift of $\frac{1}{2}$ for even p to match the asymptotic behaviors of both sides. Then we can introduce the rotated subdominant solution $\Psi_k^{(p)} = \hat{\Omega}_{-k} \Psi^{(p)}$ in sector S_k. At the origin, the basis of the solutions is made of $X_{i_1}^{(1)} \wedge \cdots \wedge X_{i_p}^{(1)}$. The highest weight vector $X_1^{(p)}$ in $V^{(p)}$ is given by $X_1^{(1)} \wedge \cdots \wedge X_p^{(1)}$. The $(r+1)$-th anti-symmetric product is isomorphic to a one-dimensional representation, which gives the determinant representation.

Given the subdominant solutions $\Psi_{i_1}^{(1)}, \ldots, \Psi_{i_{r+1}}^{(1)}$ in $V^{(1)}$, we introduce a skew-symmetric product

$$\langle \Psi_{i_1}^{(1)}, \Psi_{i_2}^{(1)}, \ldots, \Psi_{i_{r+1}}^{(1)} \rangle := \det \left(\Psi_{i_1}^{(1)}, \Psi_{i_2}^{(1)}, \ldots, \Psi_{i_{r+1}}^{(1)} \right), \tag{3.53}$$

which is independent of z and $\bar{z}$. We normalize $\Psi_k^{(1)}$ such that

$$\langle \Psi_i^{(1)}, \Psi_{i+1}^{(1)}, \ldots, \Psi_{i+r}^{(1)} \rangle = 1, \tag{3.54}$$

for any i. In Eq. (3.44), we define the Q-function $Q_i^{(p)}$ for $V^{(p)}$ by expanding $\Psi^{(p)}$ in terms of the basis of solutions around the origin $X_i^{(p)}$,

$$\Psi^{(p)} = \sum_{i=1}^{\dim V^{(p)}} Q_i^{(p)} X_i^{(p)}. \tag{3.55}$$

Expanding wedge product (3.52) in terms of the basis $X_{i_1}^{(1)} \wedge \cdots \wedge X_{i_p}^{(1)}$ around the origin, the Q-functions $Q_i^{(p)}$ can be expressed in terms of $Q_j^{(1)}$. In particular, the coefficient of the highest weight vector $X_1^{(1)} \wedge X_2^{(1)} \wedge \cdots \wedge X_p^{(1)}$ is $Q_1^{(p)}$, which is given by

$$Q_1^{(p)}(\lambda) = W_{-\frac{p-1}{2}, 1-\frac{p-1}{2}, \ldots, \frac{p-1}{2}}^{(p)}(\lambda), \tag{3.56}$$

where $W_{i_1 i_2 \ldots i_p}^{(p)}$ is the determinant of a $p \times p$ matrix with the (k, ℓ) element $Q_{k, [-i_\ell]}^{(1)}(\lambda)$ $(1 \le k, \ell \le p)$:

$$W_{i_1 i_2 \ldots i_p}^{(p)} := \det_{k, \ell} Q_{k, [-i_\ell]}^{(1)}(\lambda). \tag{3.57}$$

We then discuss the T-Q relation satisfied by the Q-functions defined above. Using Cramer's rule for determinants, we obtain the Plüker relation on the Wronskinan determinants:

$$W^{(p-1)}_{i_0 i_2 \cdots i_{p-1}} W^{(p)}_{i_1 i_2 \cdots i_p} - W^{(p-1)}_{i_1 i_2 \cdots i_{p-1}} W^{(p)}_{i_0 i_2 \cdots i_p} + W^{(p-1)}_{i_2 \cdots i_{p-1} i_p} W^{(p)}_{i_0 i_1 \cdots i_{p-1}} = 0, \qquad (3.58)$$

for $p = 1, \ldots, r+1$. For $(i_0, i_1, \ldots, i_{p-1}, i_p) = (0, 1, \ldots, p-1, p)$, (3.58) is rewritten as

$$\frac{W^{(p)}_{02 \cdots p}}{W^{(p)}_1} = \frac{W^{(p-1)}_{02 \cdots p-1}}{W^{(p-1)}_1} + \frac{W^{(p-1)}_2 W^{(p)}_0}{W^{(p)}_1 W^{(p-1)}_1} = \sum_{m=0}^{p-1} \frac{W^{(m)}_{2,\ldots,m+1} W^{(m+1)}_{0,\ldots,m}}{W^{(m)}_{1,\ldots,m} W^{(m+1)}_{1,\ldots,m+1}}, \qquad (3.59)$$

where $W^{(0)}_k = 1$ for any k. We have applied the Plücker relation several times. Let $p = r + 1$ and shift all the indexes by -1, then one obtains

$$W^{(r+1)}_{-11 \cdots r} \prod_{j=0}^{r} W^{(j)}_{0,\ldots,j-1} = \sum_{m=0}^{r} \left(\prod_{j=0}^{m-1} W^{(j)}_{0,\ldots,j-1} \right) W^{(m)}_{1,\ldots,m} W^{(m+1)}_{-1,\ldots,m-1} \left(\prod_{j=m+2}^{r+1} W^{(j)}_{0,\ldots,j-1} \right),$$
$$\qquad (3.60)$$

where we have used the normalization condition $W^{(r+1)}_{k,k+1,\ldots,k+r} = 1$.

In this equation, $W^{(p)}_{0,\ldots,p-1}$ is regarded as the $Q^{(p)}_1 (\lambda + \frac{p-1}{2} \frac{2\pi i}{hM})$, and $W^{(r+1)}_{-11 \cdots r}$ is regarded as the T-function which will be discussed in the next section. Then Eq. (3.60) becomes the form of Baxter's T-Q relation.

Suppose $\Lambda^{(k)}_i$ $(k = 1, \ldots, r)$ to be the zeros of $W^{(k)}_{0,1,\ldots,k-1}$ $(W^{(k)}_{0,1,\ldots,k-1}(\Lambda^{(k)}) = 0)$. Substituting $\Lambda = \Lambda^{(k)}_i$ into Eq. (3.60), the left-hand side and most terms of the right-hand side of Eq. (3.60) vanish, which gives the following equations for the $\Lambda^{(k)}_i$,

$$-1 = \left. \frac{W^{(k-1)}_{0,\ldots,k-2} W^{(k)}_{1,\ldots,k} W^{(k+1)}_{-1,\ldots,k-1}}{W^{(k-1)}_{1,\ldots,k-1} W^{(k)}_{-1,\ldots,k-2} W^{(k+1)}_{0,\ldots,k}} \right|_{\Lambda = \Lambda^{(k)}_i}. \qquad (3.61)$$

Setting $\lambda^{(k)}_i = \Lambda^{(k)}_i + \frac{k-1}{2} \frac{2\pi i}{hM}$, these equations become the Bethe ansatz equations (3.51). Applying other indices $(i_0, i_1, \ldots, i_{p-1}, i_p)$ in the Plücker relation (3.58), one can derive another type of T-Q relations.

We can further study the analytical structure of the Q-function $Q_1(\lambda)$ in λ. The large $|\lambda|$ asymptotic of $Q^{(1)}_1(\lambda)$ is evaluated from the definition

$$Q^{(1)}_1(\lambda) = \langle \Psi^{(1)}_0, \mathcal{X}^{(1)}_2, \ldots, \mathcal{X}^{(1)}_{r+1} \rangle \qquad (3.62)$$

in the limit $z \to 0$, which is a straightforward generalization of the definition in (1.82). From this asymptotic and requiring the analyticity in the λ-plane, the Q-functions take the infinite product form, which has zeros at the positions of the Bethe roots. We then can follow the procedure in [21–24] to introduce the counting function.[2]

---

[2] See Sects. A.3 and A.4 for the review of this procedure to rewrite the BAEs.

The BAEs finally can be rewritten as non-linear integral equations (NLIEs) of the counting function [20].

At the end of this subsection, let us note that the $A_1^{(1)}$-type modified affine Toda field equation, namely the modified sinh-Gordon equation, has been studied in [1]. From the linear problem, the authors have derived the Q-function and the NLIE of the quantum sine-Gordon model when $M > 0$, which is the first example of the massive ODE/IM correspondence.

3.2.2 T-/Y-System

We now discuss the T- and Y-system corresponding to the linear problem for the modified affine Toda field equation for $A_r^{(1)}$. First, we construct the T-system based on the Wronskian determinants. We consider the linear problem for the fundamental representation $V^{(1)}$. Let us fix the basis of the solutions of the linear problem as $\{\Psi_{-r+1}^{(1)}, \ldots, \Psi_0^{(1)}, \Psi_1^{(1)}\}$. Then $\Psi_k^{(1)}$ is expanded as

$$\Psi_k^{(1)} = (-1)^r \mathcal{T}_{1,k-2}^{[k]} \Psi_{-r+1}^{(1)} + (-1)^{r-1} \mathcal{T}_{r,k-1}^{[k-1]} \Psi_{-r+2}^{(1)} + \cdots - \mathcal{T}_{2,k-1}^{[k-1]} \Psi_0^{(1)} + \mathcal{T}_{1,k-1}^{[k-1]} \Psi_1^{(1)}, \tag{3.63}$$

where the coefficient $\mathcal{T}_{j,m}$ with $j = 1, \ldots, r$ and $m \in \mathbb{Z}$ is defined by

$$\mathcal{T}_{j,m} = \langle \Psi_{-r+1}^{(1)}, \Psi_{-r+2}^{(1)}, \ldots, \Psi_{-j+1}^{(1)}, \Psi_{-j+3}^{(1)}, \ldots, \Psi_1^{(1)}, \Psi_{m+1}^{(1)} \rangle^{[-m]}. \tag{3.64}$$

Here the superscription is defined by $f^{[m]}(\lambda) = f(\lambda + \frac{m}{2} \frac{2\pi i}{hM})$ and

$$\langle \Psi_{i_1}^{(1)}, \Psi_{i_2}^{(1)}, \ldots, \Psi_{i_{r+1}}^{(1)} \rangle^{[2]} = \langle \Psi_{i_1+1}^{(1)}, \Psi_{i_2+1}^{(1)}, \ldots, \Psi_{i_{r+1}+1}^{(1)} \rangle. \tag{3.65}$$

By the definition, one finds $\mathcal{T}_{j,-j+1} = (-1)^{j-1}$ and $\mathcal{T}_{j,m} = 0$ when $-r \leq m \leq -j, -j+2 \leq m \leq 0$. The T-function in the Baxter TQ relation (3.60) can be written as $W_{-1,1,\ldots,r}^{(r+1),[-2r]} = (-1)^r \mathcal{T}_{1,-r-2}^{[r+2]}$. From the Plücker identities

$$\langle \Psi_{j_1}^{(1)}, \Psi_{j_2}^{(1)}, \ldots, \Psi_{j_{k-2}}^{(1)}, \Psi_{j_{k-1}}^{(1)}, \Psi_{i_1}^{(1)} \rangle \langle \Psi_{j_1}^{(1)}, \Psi_{j_2}^{(1)}, \ldots, \Psi_{j_{k-2}}^{(1)}, \Psi_{i_2}^{(1)}, \Psi_{i_3}^{(1)} \rangle$$
$$- \langle \Psi_{j_1}^{(1)}, \Psi_{j_2}^{(1)}, \ldots, \Psi_{j_{k-2}}^{(1)}, \Psi_{j_{k-1}}^{(1)}, \Psi_{i_2}^{(1)} \rangle \langle \Psi_{j_1}^{(1)}, \Psi_{j_2}^{(1)}, \ldots, \Psi_{j_{k-2}}^{(1)}, \Psi_{i_1}^{(1)}, \Psi_{i_3}^{(1)} \rangle$$
$$+ \langle \Psi_{j_1}^{(1)}, \Psi_{j_2}^{(1)}, \ldots, \Psi_{j_{k-2}}^{(1)}, \Psi_{j_{k-1}}^{(1)}, \Psi_{i_3}^{(1)} \rangle \langle \Psi_{j_1}^{(1)}, \Psi_{j_2}^{(1)}, \ldots, \Psi_{j_{k-2}}^{(1)}, \Psi_{i_1}^{(1)}, \Psi_{i_2}^{(1)} \rangle = 0, \tag{3.66}$$

these $\mathcal{T}$-functions are found to satisfy the relations

$$\mathcal{T}_{j,1}^{[+]} \mathcal{T}_{1,k-1}^{[k+1]} = \mathcal{T}_{j,k}^{[k]} + \mathcal{T}_{j+1,k-1}^{[k+1]}, \quad j = 1, 2, \ldots, r - 1,$$
$$\mathcal{T}_{r,1}^{[+]} \mathcal{T}_{1,k-1}^{[k+1]} = \mathcal{T}_{r,k}^{[k]} + \mathcal{T}_{1,k-2}^{[k+2]}. \tag{3.67}$$

We expect that $\mathcal{T}$ becomes the T-function. However, $\mathcal{T}_{1,k}^{[+]} \mathcal{T}_{1,k}^{[-]}$ turns out to be

$$\mathcal{T}_{1,k}^{[+]}\mathcal{T}_{1,k}^{[-]} = \mathcal{T}_{1,k+1}\mathcal{T}_{1,k-1} + \langle \Psi_{-r+2}^{(1)}, \Psi_{-r+3}^{(1)}, \ldots, \Psi_0^{(1)}, \Psi_{k+1}^{(1)}, \Psi_{k+2}^{(1)} \rangle^{[-k-1]}, \quad (3.68)$$

where the right-hand side is not in the form of $\mathcal{T}_{a,s}$ and thus does not provide a closed system. Instead, we introduce the function $T_{a,s}$, which is designed to satisfy the T-system

$$T_{2k,m} = \langle \Psi_{-r+k+1}^{(1)}, \Psi_{-r+k+2}^{(1)}, \ldots, \Psi_{-k+1}^{(1)}, \Psi_{m+2-k}^{(1)}, \Psi_{m+3-k}^{(1)}, \ldots, \Psi_{m+k+1}^{(1)} \rangle^{[-m-1]}$$

$$T_{2k+1,m} = \langle \Psi_{-r+k+1}^{(1)}, \Psi_{-r+k+2}^{(1)}, \ldots, \Psi_{-k}^{(1)}, \Psi_{m+1-k}^{(1)}, \Psi_{m+2-k}^{(1)}, \ldots, \Psi_{m+k+1}^{(1)} \rangle^{[-m]},$$
$$(3.69)$$

where $T_{1,m} = \mathcal{T}_{1,m}$ and any other $\mathcal{T}$-functions can be expressed in terms of T-functions via the Plücker relations. We find the functions (3.69) satisfy

$$T_{a,m}^+ T_{a,m}^- = T_{a,m+1}T_{a,m-1} + T_{a-1,m}T_{a+1,m}, \quad a = 1, \ldots, r, \ m = 1, \ldots, \quad (3.70)$$

where $T_{0,m} = 1 = T_{r+1,m}$.

We can further introduce the Y-functions from the T-functions:

$$Y_{a,m} = \frac{T_{a-1,m}T_{a+1,m}}{T_{a,m+1}T_{a,m-1}}, \quad a = 1, \ldots, r. \quad (3.71)$$

From the T-system, the Y-functions are shown to satisfy

$$Y_{a,m}^+ Y_{a,m}^- = \frac{(1 + Y_{a-1,m})(1 + Y_{a+1,m})}{(1 + Y_{a,m-1}^{-1})(1 + Y_{a,m+1}^{-1})}. \quad (3.72)$$

We next consider the boundary condition of the Y-functions. It depends on the order hM of the potential and vector g in Eq. (3.24) for the solution of the modified affine Toda field equation around the origin.

We first assume that M in $p(z)$ is a generic irrational real number. The range of Stokes sector of different $\Psi_k^{(1)}(z)$, $\Psi_j^{(1)}(ze^{i2\pi\ell})$ do not coincide with each other for integers k, j, ℓ. In this case, we have an infinite number of Stokes sectors covering the z-plane and an infinite number of subdominant solutions. The T-system continues to infinity. This type of T-system is called the (A_1, A_∞)-type.

In other cases, for rational hM, the finite number of Stokes sectors defines the universal cover over the z-plane. For example $h(M + 1) = n$ with some positive integer n, there exist n independent sectors on the z-plane, where S_k and S_{n+k} completely overlap. Then the subdominant solution $\Psi_{k+n}^{(1)}$ and the subdominant solution $\Psi_k^{(1)}$ are not independent. After going around the complex z-plane once, they thus should be proportional to each other:

$$\Psi_{k+n}^{(1)}(z) \propto \Psi_k^{(1)}(ze^{-2\pi i}). \quad (3.73)$$

Then the T-system truncates up to n. The precise way of truncation depends on the monodromy around the origin, which is controlled by the parameter g.

Let us first consider the case of $g = 0$. In this case, there is no singularity at the origin. The small solutions are single-valued on the complex z-plane

$$\Psi_k^{(1)}(ze^{-2\pi i}) = \Psi_k^{(1)}(z, \lambda). \tag{3.74}$$

Then

$$\Psi_{k+n}^{(1)}(z) \propto \Psi_k^{(1)}(z). \tag{3.75}$$

Therefore we find $T_{a,n-r} = 0, a = 1, 2, \ldots, r$. This provides a natural truncation for (A_r, A_{hM-1}) T-, Y-system:

$$T_{a,m} \quad m = 0, 1, \ldots, hM, \tag{3.76}$$

$$Y_{a,m} \quad m = 1, 2, \ldots, hM - 1 \quad \text{as } h(M + 1) \in \mathbb{Z}_{>0} \text{ and } g = 0. \tag{3.77}$$

Let us consider the Y-system for the A_1 case. The Y-system becomes

$$Y_{1,m}^+ Y_{1,m}^- = \frac{1}{(1 + Y_{1,m-1}^{-1})(1 + Y_{1,m+1}^{-1})}, \quad m = 1, 2, \ldots, n - 3. \tag{3.78}$$

Replacing $Y_{1,m} \to 1/Y_m$, this Y-system reproduces the one used in Eq. (2.88). However, the analytic properties of the Y-function are much different with the Schrödinger equation, because of the singularity at $\lambda \to \pm\infty$. In a similar way to (2.92), we derive the asymptotic behavior of the Y-functions

$$\log Y_{1,s}(\lambda) \sim \begin{cases} e^\lambda m_s & \lambda \to \infty \\ e^{-\lambda} m_s & \lambda \to -\infty \end{cases}, \tag{3.79}$$

where m_s is defined by the period integral of $p(z)$ around contour γ_s.[3] Here the function $p(z)$ can be a generic polynomial. In the minimal chamber, the resulting TBA equations are

$$\log Y_{1,s} = 2m_s \cosh \lambda - K * \log\left((1 + Y_{1,s-1}^{-1})(1 + Y_{1,s+1}^{-1})\right), \tag{3.80}$$

which are the massive generalization of the TBA equations (2.97). This set of TBA equations has been used to compute the minimal area of the worldsheet ending on the null polygon Wilson loop in the AdS_3 boundary, which is dual to the planar limit of the scattering amplitude of $\mathcal{N} = 4$ SYM at strong coupling [16, 17].[4]

For $g \neq 0$, the subdominant solution is no longer single-valued on the z plane, because of the non-trivial monodromy at $z = 0$. Here we discuss the $A_1^{(1)}$ case. Let $\{\Psi_0^{(1)}, \Psi_1^{(1)}\}$ be a basis of the solution. The monodromy matrix $\Omega(\lambda)$ is defined by

---

[3] See [16] for the derivation of the asymptotics.

[4] See also [16, 25, 26] the discussion of the wall-crossing of the TBA equations. The TBA equations for the non-planar scattering amplitude at strong coupling can be found in [27].

$$\begin{pmatrix} \Psi_1^{(1)} \\ \Psi_0^{(1)} \end{pmatrix} (ze^{2\pi i}, \zeta) = \Omega(\zeta) \begin{pmatrix} \Psi_1^{(1)} \\ \Psi_0^{(1)} \end{pmatrix} (z, \zeta), \tag{3.81}$$

where $\det[\Omega(\lambda)] = 1$ in our normalization. We can expand these subdominant solutions using the basis $\mathcal{X}_i^{(1)}$ which has monodromy $e^{-2\pi i \beta g \cdot h_i^{(1)}}$. We express $\mathrm{Tr}[\Omega]$ as

$$\mathrm{tr}[\Omega(\lambda)] = 2 \cos(2\pi \beta g \cdot h_1^{(1)}), \tag{3.82}$$

where $h_1^{(1)} + h_2^{(1)} = 0$ is used. We then introduce the proportionality coefficient in (3.73) with $k = 0$

$$\Psi_n^{(1)}(z, \lambda) = B(\lambda) \Psi_0^{(1)}(ze^{-2\pi i}, \lambda), \tag{3.83}$$

from which one finds

$$\Psi_{n+1}^{(1)}(z, \lambda) = B(\lambda)^{-1} \Psi_1^{(1)}(ze^{-2\pi i}, \lambda). \tag{3.84}$$

Using the quasi-periodic condition of the Q-function (3.45), one finds that $B(\lambda) = e^{i\pi}$, which leads to

$$\begin{pmatrix} \Psi_{n+1}^{(1)} \\ \Psi_n^{(1)} \end{pmatrix} (z, \lambda) = -\Omega^{-1}(\lambda) \begin{pmatrix} \Psi_1^{(1)} \\ \Psi_0^{(1)} \end{pmatrix} (z, \lambda). \tag{3.85}$$

We thus can express $\Omega(\lambda)$ in terms of the Wronskinan of subdominant solutions such as

$$\mathrm{Tr}\Omega = T_{1,n-2}^{[n]} - T_{1,n}^{[n]}. \tag{3.86}$$

We thus can express $T_{1,n}$ in terms of $T_{1,n-2}$, and the monodromy matrix. This provides a natural truncation for the T-system

$$T_{1,m} \quad a = 1, 2, \dots, r, \quad m = 1, \dots, n - 1. \tag{3.87}$$

Since $Y_{a,n-1}$ contains $T_{a,n}$, it is natural to stop at $Y_{a,n-2}$. Their functional relations are

$$Y_{1,m}^+ Y_{1,m}^- = \frac{1}{(1 + Y_{a,m-1}^{-1})} \frac{1}{(1 + Y_{a,m+1}^{-1})}, \quad m = 1, \dots, n - 2. \tag{3.88}$$

Note that when $m = n - 2$, there exists $Y_{a,n-1}$ on the right-hand side of the equations. To obtain a closed Y-system, we need to introduce a new Y-function

$$\bar{Y} = -\frac{1}{T_{1,n-2}}, \tag{3.89}$$

which satisfies $Y_{1,n-1} = -\frac{\bar{Y}}{T_{1,n}}$ and

$$\bar{Y}^{+}\bar{Y}^{-} = \frac{1}{(1 + Y_{a,n-2}^{-1})}. \tag{3.90}$$

Expressing $Y_{1,n}$ and $T_{1,n}$ in terms of the Y-functions, one finds

$$Y_{1,n-2}^{+}Y_{1,n-2}^{-} = \frac{1}{(1 + Y_{a,n-3}^{-1})} \frac{1}{\left(1 + \bar{Y}^{-2} + 2\cos(2\pi\beta g \cdot h_1^{(1)})\bar{Y}^{-1}\right)}. \tag{3.91}$$

In summary, we obtain a closed Y-system [1]

$$Y_{1,m}^{+}Y_{1,m}^{-} = \frac{1}{(1 + Y_{a,m-1}^{-1})} \frac{1}{(1 + Y_{a,m+1}^{-1})}, \, m = 1, \dots, n-3,$$

$$Y_{1,n-2}^{+}Y_{1,n-2}^{-} = \frac{1}{(1 + Y_{a,n-3}^{-1})} \frac{1}{\left(1 + \bar{Y}^{-2} + 2\cos(2\pi\beta g \cdot h_1^{(1)})\bar{Y}^{-1}\right)}, \tag{3.92}$$

$$\bar{Y}^{+}\bar{Y}^{-} = \frac{1}{(1 + Y_{a,n-2}^{-1})}.$$

We observe that the Y-system for the $A_1^{(1)}$ modified affine Toda field equation with monomial $p(z)$ is the same as the Y-system (1.75) for the Schrödinger equation with the centrifugal potential term. However, the difference arises in the asymptotic behavior of the Y-function at $\lambda \to \pm\infty$, which leads to the different driving terms in the TBA equation. For the linear problem associated with the modified affine Toda field equation, we have the massive TBA equation with the driving term of the form $m \cosh \theta$, while for massless TBA, it is of the form me^{θ}. These TBA equations have been used to compute the minimal area of the worldsheet ending on the periodic null polygon Wilson line in the AdS$_3$ boundary, which is dual to the form factor of $\mathcal{N} = 4$ SYM at strong coupling [28, 29].

In the case of $A_r^{((1)}$ with $r \geq 2$, the generalization of the monodromy (3.86) is much more complicated. The systematic way to truncate the Y-system is not yet known. See [29, 30] for some attempts.

3.3 Light-Cone Limit and the ODE/IM Correspondence

We consider the light-cone limit of the linear problem associated with the modified affine Toda field equation and show how to reproduce the massless ODE/IM correspondence from this limit.

The linear problem (3.20) associated with the affine Lie algebra $\hat{\mathfrak{g}}$ is made of two linear differential equations. We can consider the light-cone limit $\bar{z} \to 0$ with fixed z, where we regard $\bar{z}$ and z as independent variables. Here the field $\phi(z, \bar{z})$ has been assumed to behave as (3.24) in this limit. After taking the light-cone limit, only the holomorphic part of the linear problem $\mathcal{D}\psi = 0$ remains. We then take the conformal limit $z \to 0$ and $s \to 0$ with the following variables:

$$x = (me^\lambda)^{1/(M+1)} z, \quad E = s^{hM} (me^\lambda)^{hM/(M+1)}, \tag{3.93}$$

which are kept finite. $p(z, s)$ scales as $p(z, s) = (me^\lambda)^{\frac{-hM}{M+1}} p(x, E)$, with $p(x, E) \equiv x^{hM} - E$.

For an affine Lie algebra $\hat{\mathfrak{g}}$ with rank r and its representation V, after taking the light-cone and the conformal limit, the holomorphic linear problem leads to the linear differential system

$$\mathcal{L}_{\hat{\mathfrak{g}}}(x, E, l; \zeta) \; \Psi(x, E, l) = \left[\frac{d}{dx} + A_{\hat{\mathfrak{g}}} \right] \Psi(x, E, l) = 0. \tag{3.94}$$

Ψ is a V-valued function of x and $A_{\hat{\mathfrak{g}}}$ is the $\hat{\mathfrak{g}}$-valued matrix defined by

$$A_{\hat{\mathfrak{g}}} := -\frac{1}{x} \sum_{a=1}^{r} l_a (\alpha_a^\vee \cdot H) + \sum_{a=1}^{r} \sqrt{n_a^\vee} E_{\alpha_a} + \sqrt{n_0^\vee} \, p(x, E) \, \zeta \, E_{\alpha_0}, \tag{3.95}$$

where

$$l_a := -\omega_a \cdot g. \tag{3.96}$$

The parameters $l = (l_1, \ldots, l_r)$ characterize the monodromy properties of the solution around $x = 0$. g is an r-dimensional vector that satisfies the same condition as in (3.24):

$$1 + \alpha_a \cdot g > 0, \quad a = 0, \ldots, r. \tag{3.97}$$

ζ is fixed to be $+1$ or -1 so that the subdominant solution (3.111) along the real axis exists.

For the linear problem (3.94) in the fundamental representation $V^{(1)}$ with the highest weight ω_1, one can derive the ODE for the highest weight component of ψ [4]. Let us consider the case of $A_1^{(1)}$ in the fundamental representation $V^{(1)}$. Let e_1, e_2 be the orthonormal basis of $V^{(1)}$ with $\Psi = \psi_1 e_1 + \psi_2 e_2$. The linear problem takes the form

$$\left[\begin{pmatrix} \frac{d}{dx} - \frac{l_1}{x} & 0 \\ 0 & \frac{d}{dx} + \frac{l_1}{x} \end{pmatrix} + \begin{pmatrix} 0 & 1 \\ p(x, E) & 0 \end{pmatrix} \right] \begin{pmatrix} \psi_1 \\ \psi_2 \end{pmatrix} = 0. \tag{3.98}$$

Eliminating ψ_2, one finds ψ_1 satisfies the Schrödinger equation with angular momentum and the monomial potential [31]:

$$\left[-\frac{d^2}{dx^2} + \frac{l_1(l_1 - 1)}{x^2} + x^{2M} - E \right] \psi_1 = 0. \tag{3.99}$$

We can find the higher-order ODEs for other affine Lie algebras. Here we write down the ODE for $A_r^{(1)}$, $D_r^{(1)}$, $A_{2r-1}^{(2)}$ and $D_{r+1}^{(2)}$. For other cases see [4] or [11].

- For the fundamental representation $V^{(1)}$ of $A_r^{(1)}$, the weight vectors are $h_1, \ldots, h_{r+1}$, where $h_1 = \omega_1$ and $h_i - h_{i+1} = \alpha_i$ ($i = 1, \ldots, r$). The highest root is given by $\theta = \alpha_1 + \cdots + \alpha_{r+1}$ and the Coxeter number is $h = r + 1$. The representation of the generators is given by $E_{\alpha_i} = e_{i,i+1}$ and $E_{\alpha_0} = e_{r+1,1}$ and $E_{-\alpha} = E_\alpha^T$, where $e_{a,b}$ denotes the order $\dim V$ matrix with (i, j) element $\delta_{ai}\delta_{bj}$. The linear problem (3.94) is solved for ψ_1 as

$$\left(D(h_{r+1}) \cdots D(h_1) - (-1)^h p(x, E)\right) \psi(x) = 0. \qquad (3.100)$$

Here $D(h) \equiv \partial + \beta \frac{h \cdot g}{x}$.

- For $D_r^{(1)}$, the representation $V^{(1)}$ is $(2r)$-dimensional and the weight vectors $h_1, \ldots, h_{2r}$ satisfy $h_1 = \omega_1$, $h_i - h_{i+1} = \alpha_i$ ($i = 1, \ldots, r - 1$) and $h_{2r+1-i} = -h_i$ ($i = 1, \ldots, r$) and its representation is $E_{\alpha_i} = e_{i,i+1} + e_{2r-i,2r+1-i}$, $E_{\alpha_r} = e_{r-1,r+1} + e_{r,r+2}$ and $E_{\alpha_0} = e_{2r-1,1} + e_{2r,2}$. The highest root is $\theta = \alpha_1 + 2\alpha_2 + \cdots + 2\alpha_{r-2} + \alpha_{r-1} + \alpha_r$ and $h = 2r - 2$. Solving the linear problem for ψ_1, we obtain the higher–order differential equation including the pseudo-differential operator ∂^{-1}:

$$\left[D(-h_1) \ldots D(-h_r)\partial^{-1} D(h_r) \ldots D(h_1) - 2^{r-1}\sqrt{p(x, E)}\partial\sqrt{p(x, E)}\right] \psi_1(x) = 0. \qquad (3.101)$$

- For the twisted affine Lie algebra $A_{2r-1}^{(2)}$ which is dual to the untwisted affine Lie algebra $B_r^{(1)}$, the highest root is given by $\theta = \alpha_1 + 2\alpha_2 + \cdots + 2\alpha_{r-1} + \alpha_r$ ($h = 2r - 2$) and the $(2r + 1)$-dimensional representation has the weight vectors h_i with $h_i - h_{i+1} = \alpha_i$ ($i = 1, \ldots, r - 1$), $h_r = \frac{1}{2}\alpha_r$ and $h_{2r+1-i} = -h_i$ ($i = 1, \ldots, r$). From the matrix representation $E_{\alpha_i} = e_{i,i+1} + e_{2r-i,2r+1-i}$, $E_{\alpha_r} = e_{r,r+1}$ and $E_{\alpha_0} = e_{2r,2} + e_{2r-1,1}$, one obtains the ODE for ψ_1:

$$\left[D(-h_1) \ldots D(-h_r)D(h_r) \ldots D(h_1) + 2^{r-1}\sqrt{p(x, E)}\partial\sqrt{p(x, E)}\right] \psi_1(x) = 0. \qquad (3.102)$$

- For the twisted affine Lie algebra $D_{r+1}^{(2)} = (C_r^{(1)})^\vee$, the highest root is $\theta = \alpha_1 + \cdots + \alpha_r$ ($h = r$) and the $(2r + 2)$-dimensional representation has the weights h_i with the highest weight ω_1. From the matrix representation $E_{\alpha_i} = e_{i,i+1} + e_{2r+2-i,2r+3-i}$ ($i = 1, \ldots, r - 1$), $E_{\alpha_r} = \sqrt{2}(e_{r,r+1} + e_{r+1,r+3})$ and $E_{\alpha_0} = \sqrt{2}(e_{r+2,1} + e_{2r+2,r+2})$, we obtain the ODE for ψ_1:

$$\left[D(-h_1) \ldots D(-h_{r+1})\partial D(h_r) \ldots D(h_1) - 2^{r+1}\sqrt{p(x, E)}\partial^{-1}\sqrt{p(x, E)}\right] \psi_1(x) = 0. \qquad (3.103)$$

Let us study the asymptotic solution of the linear problem (3.94). As in (3.25) under the rotation $(x, E) \to (\omega^k x, \Omega^k E)$ with $\Omega = e^{\frac{2\pi i M}{M+1}}$ and $\omega = e^{\frac{2\pi i}{h(M+1)}}$, the linear differential operator $\mathcal{L}_{\hat{\mathfrak{g}}}$ transforms as

$$\omega^k \mathcal{L}_{\hat{\mathfrak{g}}}(\omega^k x, \Omega^k E, l; \zeta) = \omega^{k\rho^\vee \cdot H} \mathcal{L}_{\hat{\mathfrak{g}}}(x, E, l; \zeta e^{2\pi ik})\omega^{-k\rho^\vee \cdot H}. \tag{3.104}$$

Let $\Psi(x, E)$ be the solution of the linear problem $\mathcal{L}_{\hat{\mathfrak{g}}}(x, E, l; \zeta)$. Then

$$\Psi_{[k]}(x, E) := \omega^{-k\rho^\vee \cdot H} \Psi(\omega^k x, \Omega^k E) \tag{3.105}$$

satisfies the linear problem for $\mathcal{L}_{\hat{\mathfrak{g}}}(x, E, l; \zeta)$ evaluated by replacing $E_{\alpha_0} \to e^{2\pi ik} E_{\alpha_0}$:

$$\mathcal{L}_{\hat{\mathfrak{g}}}(x, E, l; \zeta e^{2\pi ik})\Psi_{[k]}(x, E) = 0. \tag{3.106}$$

We define the representation $V_{[k]}$, where the subscript $[k]$ means that the generator E_{α_0} acts on V as $e^{2\pi ik} E_{\alpha_0}$. If Ψ is a solution of the linear problem on V, then $\Psi_{[k]}$ becomes a solution on $V_{[k]}$.

We discuss the solution of the linear problem at infinity using the WKB approximation. Consider the gauge transformation: $\Psi(x) \to \Psi'(x) = U(x)\Psi(x), \mathcal{L}_{\hat{\mathfrak{g}}} \to \mathcal{L}'_{\hat{\mathfrak{g}}} = U(x)\mathcal{L}_{\hat{\mathfrak{g}}}U^{-1}(x)$, where $U(x) = \exp\left(\log\left(p(x, E)\right)^{1/h} \rho^\vee \cdot H\right)$. The linear operator becomes

$$\mathcal{L}'_{\hat{\mathfrak{g}}} = \frac{d}{dx} - \frac{1}{x}\sum_{a=1}^{r} l_a \, \alpha_a \cdot H - \frac{1}{h}\frac{d\log p(x, E)}{dx}\rho^\vee \cdot H + p(x, E)^{1/h}\Lambda_+, \tag{3.107}$$

where

$$\Lambda_+ = \sum_{a=1}^{r} \sqrt{n_a^\vee} \, E_{\alpha_a} + \sqrt{n_0^\vee} \, \zeta E_{\alpha_0}. \tag{3.108}$$

At large x, the $O(1/x)$ terms in $\mathcal{L}'_{\hat{\mathfrak{g}}}$ can be ignored. Defining ν_i and $\boldsymbol{\nu}_i$ ($i = 1, \ldots, \dim V$) as the eigenvalues and the eigenvectors of Λ_+, the WKB solution $\Psi^{\text{WKB}}(x, E)$ is given by

$$\Psi^{\text{WKB}}(x, E) = \sum_{i=1}^{\dim V} C_i \exp\left(-\nu_i \int^x (p(x', E))^{1/h} dx' - \frac{1}{h}\log(p(x, E))\rho^\vee \cdot H\right)\boldsymbol{\nu}_i, \tag{3.109}$$

where C_i are constants. We denote $\Psi(x, E)$ the subdominant solution along the positive real axis. $\Psi(x, E)$ decays fastest in the sector $\mathcal{S}_0 := \{x \in \mathbf{C} \mid |\arg x| < \frac{\pi}{h(M+1)}\}$ and has the asymptotic behaviour for large x:

$$\Psi(x, E, l) \sim C \exp\left(-\nu \int^x (p(x', E))^{1/h} dx'\right)\exp\left(-\frac{1}{h}\log p(x, E)\,\rho^\vee \cdot H\right)\boldsymbol{\nu}, \tag{3.110}$$

$$\sim Cx^{-M\rho^\vee \cdot H} \exp\left(-\frac{\nu}{M+1} x^{M+1}\right) \boldsymbol{v}, \qquad x \to \infty, \tag{3.111}$$

where C is a constant. ν is the eigenvalue of generator Λ_+ with the largest real value and $\boldsymbol{v}$ the associated eigenvector. The Symanzik rotation of the solution $\Psi_{[k]}(x, E, l)$ ($k \in \mathbf{Z}$) becomes the subdominant solution in the sector $\mathcal{S}_k$, which is defined by

$$\mathcal{S}_k : \quad \left| \arg x + \frac{2\pi k}{h(M+1)} \right| < \frac{\pi}{h(M+1)}. \tag{3.112}$$

We next discuss the solution of the linear problem around $x = 0$. We find a basis of power series solutions $\mathcal{X}_i(x, E, l)$ ($i = 1, \ldots, \dim V$) around $x = 0$:

$$\mathcal{X}_i(x, E, l) = x^{-h_i \cdot g} \boldsymbol{e}_i + \cdots, \qquad x \to 0. \tag{3.113}$$

$\mathcal{X}_i(x, E, l)$ has a monodromy $e^{-2\pi i h_i \cdot g}$ around the origin. Here $\mathcal{X}_i(x, E, l)$ transforms as

$$\mathcal{X}_{i[k]}(x, E, l) = \omega^{-kh_i \cdot (\rho^\vee + g)} \mathcal{X}_i(x, E, l). \tag{3.114}$$

Using this basis, the subdominant solution $\Psi(x, E, l)$ can be expanded as

$$\Psi(x, E, l) = \sum_{i=1}^{\dim V} Q_i(E, l)\, \mathcal{X}_i(x, E, l). \tag{3.115}$$

Here the coefficient $Q_i(E, l)$ defines the generalized Q-functions, where the zeros of the $Q_i(E, l)$ would be characterized the Bethe ansatz equations. One can calculate the zeros of Q_i from the Wronskian:

$$Q_i(E, l) = \frac{\langle \mathcal{X}_1, \ldots, \Psi, \ldots, \mathcal{X}_{\dim V} \rangle}{\langle \mathcal{X}_1, \ldots, \mathcal{X}_{\dim V} \rangle}, \tag{3.116}$$

by substituting Eqs. (3.113) and (3.110) into this. Practically, it is hard to compute zeros of Q_i in (3.116) numerically for the representation with large dimensions. A better approach to computing the zeros has been studied in [11], where one introduces the dual basis of the solutions $\mathcal{X}_i^*$ with $\langle \mathcal{X}_i^*, \mathcal{X}_j \rangle = \delta_{ij}$. Here $\langle\, ,\, \rangle$ is the inner product $V^* \times V \to \mathbb{C}$. Then Q_i is calculated as

$$Q_i(E, l) = \langle \mathcal{X}_i^*, \Psi \rangle. \tag{3.117}$$

Since the r.h.s. is independent of x, we can choose x as an arbitrary real number. For sufficiently large x, the dominant contribution to (3.117) is from the lowest component of Ψ, which becomes divergent for large x. Then the zeros of $Q_i(E, l)$ are determined by the zeros of the lowest component of $\mathcal{X}_i(x, E, l)$ for fixed x. For the A_1 case, the linear problem is self-dual. One finds the zeros of the solutions $\mathcal{X}_i$ define those of the Q-function, which has been observed already in Chap. 1.

In general, we can identify the coefficients $Q_1(E, l)$ in the fundamental representation $V^{(a)}$ as the Q-functions $Q^{(a)}$ of the integrable model characterized by the BAEs. The functional relations satisfying the Q-functions including T-Q relations are very similar to those for the massive case. For details, see [9]. Using the ψ-system, we obtain the Bethe ansatz equations:

$$\prod_{b=1}^{r} \Omega^{C_{ab}\gamma_b/2} \left. \frac{Q^{(b)}_{[C_{ab}/2]}}{Q^{(b)}_{[-C_{ab}/2]}} \right|_{E_i^{(a)}} = -1, \qquad i = 0, 1, \ldots, \tag{3.118}$$

where

$$\gamma_a := \frac{2}{hM} \left(l_a - \omega_a \cdot \rho^\vee \right) \tag{3.119}$$

is regarded as the twist parameter in the integrable models (see Sects. A.3 and A.4). The relations have been tested numerically for various affine Lie algebras and their representations [11].

3.4 Summary

In this chapter, we have reviewed the massive ODE/IM correspondence associated with the affine Toda field equation modified by the conformal transformation. This correspondence describes the relation between a set of linear differential equations and the massive integrable models. The connection coefficients between the subdominant solutions and the solutions around the origin define the Q-functions. The ψ-system for the solutions of the linear problem imposes the conditions for the Q-functions, from which one derives the Bethe ansatz equations. From the Wronskian relations, we have also derived the T-/Y-system of the massive integrable QFT for $A_r^{(1)}$-type modified affine Toda field equation. These systems have the same forms as the massless integrable models but with different analytic and asymptotic properties. Taking the light-cone limit, one has the single linear problem which provides the ODE of the massless ODE/IM correspondence. We present the diagram in Fig. 3.1 to summarize.

In [20], we have derived the non-linear integral equations (NLIEs) and the effective central charge of the massive BAEs for the $A_r^{(1)}$-type modified affine Toda field equation. We have found the effective central charge in the UV limit is identified as that of the non-unitary WA_r minimal model when the solution has trivial monodromy. The NLIEs have been generalized to polynomial potential $p(z)$ for $A_1^{(1)}$ type modified affine Toda field equation, which is applied to study the scattering amplitude in the

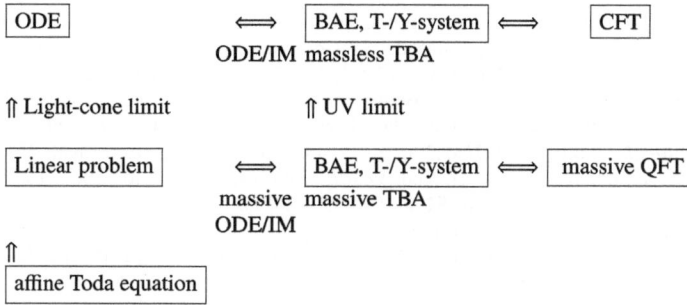

Fig. 3.1 The relation between massive and massless ODE/IM correspondences

strong coupling in the $\mathcal{N} = 4$ super Yang–Mills theory [32]. In [11], the massless ODE/IM correspondence for generic untwisted affine Lie algebras have been tested numerically. More recently, the ODE/IM correspondence has been generalized to the affine Toda field equation of an affine Lie superalgebra [33].

References

1. S.L. Lukyanov, A.B. Zamolodchikov, Quantum Sine(h)-Gordon model and classical integrable equations. JHEP **07**, 008 (2010). https://doi.org/10.1007/JHEP07(2010)008, arXiv:1003.5333 [math-ph]
2. V.G. Drinfeld, V.V. Sokolov, Lie algebras and equations of Korteweg-de Vries typeJ. Sov. Math. **30**, 1975–2036 (1984). https://doi.org/10.1007/BF02105860
3. P. Dorey, S. Faldella, S. Negro, R. Tateo, The Bethe Ansatz and the Tzitzeica-Bullough-Dodd equation. Phil. Trans. Roy. Soc. Lond. **A371**, 20120052 (2013). https://doi.org/10.1098/rsta.2012.0052, arXiv:1209.5517 [math-ph
4. K. Ito, C. Locke, ODE/IM correspondence and modified affine Toda field equations. Nucl. Phys. **B885**, 600–619 (2014). https://doi.org/10.1016/j.nuclphysb.2014.06.007, arXiv:1312.6759 [hep-th]
5. P. Adamopoulou, C. Dunning, Bethe Ansatz equations for the classical $A_n^{(1)}$ affine Toda field theories. J. Phys. **A47**, 205205 (2014). https://doi.org/10.1088/1751-8113/47/20/205205, arXiv:1401.1187 [math-ph]
6. K. Ito, C. Locke, ODE/IM correspondence and Bethe ansatz for affine Toda field equations. Nucl. Phys. **B896**, 763–778 (2015). https://doi.org/10.1016/j.nuclphysb.2015.05.016, arXiv:1502.00906 [hep-th]
7. P. Dorey, C. Dunning, D. Masoero, J. Suzuki, R. Tateo, Pseudo-differential equations, and the Bethe ansatz for the classical Lie algebras. Nucl. Phys. **B772**, 249–289 (2007). https://doi.org/10.1016/j.nuclphysb.2007.02.029, arXiv:hep-th/0612298 [hep-th]
8. J. Sun, Polynomial relations for q-characters via the ODE/IM correspondence. SIGMA**8**, 028 (2012). https://doi.org/10.3842/SIGMA.2012.028, arXiv:1201.1614 [math.QA]

9. D. Masoero, A. Raimondo, D. Valeri, Bethe Ansatz and the spectral theory of affine lie algebra–valued connections I. The simply–laced case. Commun. Math. Phys. **344**(3), 719–750 (2016). https://doi.org/10.1007/s00220-016-2643-6, arXiv:1501.07421 [math-ph]
10. D. Masoero, A. Raimondo, D. Valeri, Bethe Ansatz and the spectral theory of affine lie algebra–valued connections II: the non simply–laced case. Commun. Math. Phys. **349**(3), 1063–1105 (2017). https://doi.org/10.1007/s00220-016-2744-2, arXiv:1511.00895 [math-ph]
11. K. Ito, T. Kondo, K. Kuroda, H. Shu, ODE/IM correspondence for affine Lie algebras: a numerical approach. J. Phys. A **54**(4), 044001 (2021). https://doi.org/10.1088/1751-8121/abd21e, arXiv:2004.09856 [hep-th]
12. H.J. de Vega, M. Karowski, Exact Bethe Ansatz solution of 0(2n) symmetric theories. Nucl. Phys.**B280**, 225–254 (1987). https://doi.org/10.1016/0550-3213(87)90146-5
13. B.A. Burrington, P. Gao, Minimal surfaces in AdS space and integrable systems. JHEP **04**, 060 (2010). https://doi.org/10.1007/JHEP04(2010)060, arXiv:0911.4551 [hep-th]
14. L.F. Alday, J.M. Maldacena, Gluon scattering amplitudes at strong coupling. JHEP **06**, 064 (2007). https://doi.org/10.1088/1126-6708/2007/06/064, arXiv:0705.0303 [hep-th]
15. L.F. Alday, D. Gaiotto, J. Maldacena, Thermodynamic Bubble Ansatz. JHEP **09**. https://doi.org/10.1007/JHEP09(2011)032, arXiv:0911.4708 [hep-th]
16. L.F. Alday, J. Maldacena, A. Sever, P. Vieira, Y-system for scattering amplitudes. J. Phys. A **43**, 485401 (2010). https://doi.org/10.1088/1751-8113/43/48/485401, arXiv:1002.2459 [hep-th]
17. Y. Hatsuda, K. Ito, K. Sakai, Y. Satoh, Thermodynamic Bethe Ansatz equations for minimal surfaces in AdS_3. JHEP **04**, 108 (2010). https://doi.org/10.1007/JHEP04(2010)108, arXiv:1002.2941 [hep-th]
18. D. Gaiotto, G.W. Moore, A. Neitzke, Wall-crossing, Hitchin systems, and the WKB approximation. Adv. Math. **234**, 239–403 (2013). https://doi.org/10.1016/j.aim.2012.09.027, arXiv:0907.3987 [hep-th]
19. W. Fulton, J. Harris, *Representation Theory: A First Course*. Graduate texts in mathematics. (Springer, 1991). https://books.google.co.jp/books?id=UZWPQgAACAAJ
20. K. Ito, H. Shu, Massive ODE/IM correspondence and non-linear integral equations for $A_r^{(1)}$-type modified affine Toda field equations. J. Phys. A **51**(38), 385401 (2018). https://doi.org/10.1088/1751-8121/aad63f, arXiv:1805.08062 [hep-th]
21. C. Destri, H.J. de Vega, New thermodynamic Bethe ansatz equations without strings. Phys. Rev. Lett. **69**, 2313–2317 (1992). https://doi.org/10.1103/PhysRevLett.69.2313
22. C. Destri, H.J. De Vega, Unified approach to thermodynamic Bethe Ansatz and finite size corrections for lattice models and field theories. Nucl. Phys. **B438**, 413–454 (1995). https://doi.org/10.1016/0550-3213(94)00547-R, arXiv:hep-th/9407117 [hep-th]
23. C. Destri, H.J. de Vega, Nonlinear integral equation and excited states scaling functions in the sine-Gordon model. Nucl. Phys. B **504**, 621–664 (1997). https://doi.org/10.1016/S0550-3213(97)00468-9, arXiv:hep-th/9701107
24. P. Zinn-Justin, Nonlinear integral equations for complex affine Toda models associated to simply laced Lie algebras. J. Phys. A **31**, 6747–6770 (1998). https://doi.org/10.1088/0305-4470/31/31/019, arXiv:hep-th/9712222
25. J. Toledo, *Exact results in QFT: minimal areas and maximal couplings*. Ph.D. thesis, University of Waterloo, https://uwspace.uwaterloo.ca/handle/10012/10841 (2016)
26. J. Toledo, Notes on wall-crossing. Unpublished (2010)
27. H. Ouyang, H. Shu, TBA-like equations for non-planar scattering amplitude/Wilson lines duality at strong coupling. JHEP **05**, 099 (2022). https://doi.org/10.1007/JHEP05(2022)099, arXiv:2202.10700 [hep-th]
28. J. Maldacena, A. Zhiboedov, Form factors at strong coupling via a Y-system. JHEP**11**, 104 (2010). https://doi.org/10.1007/JHEP11(2010)104, arXiv:1009.1139 [hep-th]
29. Z. Gao, G. Yang, Y-system for form factors at strong coupling in AdS_5 and with multi-operator insertions in AdS_3. JHEP **06**, 105 (2013). https://doi.org/10.1007/JHEP06(2013)105, arXiv:1303.2668 [hep-th]
30. K. Ito, H. Shu, ODE/IM correspondence for modified $B_2^{(1)}$ affine Toda field equation. Nucl. Phys. B **916**, 414–429 (2017). https://doi.org/10.1016/j.nuclphysb.2017.01.009, arXiv:1605.04668 [hep-th]

31. V.V. Bazhanov, S.L. Lukyanov, A.B. Zamolodchikov, Spectral determinants for Schrodinger equation and Q operators of conformal field theory. J. Statist. Phys. **102**, 567–576 (2001). https://doi.org/10.1023/A:1004838616921, arXiv:hep-th/9812247
32. D. Fioravanti, M. Rossi, H. Shu, QQ-system and non-linear integral equations for scattering amplitudes at strong coupling. JHEP **12**, 086 (2020). https://doi.org/10.1007/JHEP12(2020)086, arXiv:2004.10722 [hep-th]
33. K. Ito, M. Zhu, ODE/IM correspondence and supersymmetric affine Toda field equations. Nucl. Phys. B **985**, 116004 (2022). https://doi.org/10.1016/j.nuclphysb.2022.116004, arXiv:2206.08024 [hep-th]

Appendix
Integrable Models and Functional Relations

In Chap. 1 of the main text, we discussed the relation between the Schrödinger equation with a monomial potential and the spin 1/2 XXZ spin-chain in the conformal limit. The model is also described by the two-dimensional conformal field theory (CFT). We have derived the T-system and Y-system from the Wronskians of the subdominant solutions of the ODE. These functional relations also appear in the integrable models. Furthermore one can derive the integral equations from these functional relations called the TBA equations. The TBA equations provide a useful tool to study the Y-functions analytically and numerically, which play an important role in the main text of this book.

In this Appendix, we review some basic notions in conformal field theory and integrable models which are used in this book. We emphasize the approach using the functional relations and the integral equations that appear in the integrable models. We also discuss the relation between the T-function and the integrals of motion in CFT. More details of the topic described here can be found in [1, 2] for example.

A.1 Two-Dimensional CFT

CFT is a quantum field theory that is invariant under conformal transformation. In two-dimensional spacetime, the conformal symmetry group has infinitely many generators, constraining the field theory. The dynamics of the theory are determined by a set of operators and their correlation functions. The representations of conformal symmetry organize a space of operators. In particular, the operator product expansions (OPEs) of the primary fields and the bootstrap structure of the correlation functions determine the theory, where we do not need to know the Lagrangian of the theory, which sometimes cannot be determined. Here we deal with two-dimensional CFT.

© The Author(s), under exclusive licence to Springer Nature Singapore Pte Ltd. 2025
K. Ito and H. Shu, *ODE/IM Correspondence and Quantum Periods*, SpringerBriefs in Mathematical Physics 51, https://doi.org/10.1007/978-981-96-0499-9

Let $(z, \bar{z})$ be the complex coordinates of two-dimensional spacetime with Euclidean signature. The infinitesimal conformal transformation $z \to w = z + \epsilon_n z^{n+1}$, $\bar{z} \to \bar{z} + \bar{\epsilon}_n \bar{z}^{n+1}$ $(n \in \mathbb{Z})$, which is realized by the Virasoro generators L_n, $\bar{L}_n$ on the Hilbert space of the theory. One defines the energy–momentum tensor whose holomorphic component is $T(z) = \sum_{n=-\infty}^{\infty} L_n z^{-n-2}$. The anti-holomorphic part $\bar{T}(\bar{z}) = \sum_{n=-\infty}^{\infty} \bar{L}_n \bar{z}^{-n-2}$ is also defined similarly. Here we discuss the holomorphic part only. The structure of the conformal symmetry is understood through the OPE of $T(z)$ and $T(w)$, which is given by

$$T(z)T(w) = \frac{c/2}{(z-w)^4} + \frac{2T(w)}{(z-w)^2} + \frac{\partial T(w)}{z-w} + \text{regular}. \qquad (A.1)$$

This can be translated into the commutation relations for L_n called the Virasoro algebra:

$$[L_m, L_n] = (m-n)L_{m+n} + \frac{c}{12}(m^3 - m)\delta_{m+n,0}. \qquad (A.2)$$

Here c is the central charge, which takes a real number. The contents of the operators of a CFT are determined by the primary fields $\phi_i(z, \bar{z})$ and their descendants which are obtained by the act of Virasoro generators on ϕ_i. In CFT, their OPEs are mostly interesting objects. The OPE of $T(z)$ and $\phi(w)$ is given by

$$T(z)\phi(w) = \frac{\Delta\phi(w)}{(z-w)^2} + \frac{\partial\phi(w)}{z-w} + \text{regular}, \qquad (A.3)$$

where Δ is called the conformal dimension of ϕ. The state corresponding to a primary field defines a highest-weight representation of the Virasoro algebra. For some particular values of c and Δ, the representation contains singular vectors, which are removed to construct the irreducible representation. Such a representation is called the degenerate representation.

The **minimal model** contains only the primary fields that belong to the degenerate representations of the Virasoro algebra [3]. The minimal model $M_{p,q}$ is labeled by two coprime positive integers p, q. Its central charge is given by

$$c = 1 - \frac{6(p-q)^2}{pq}. \qquad (A.4)$$

The primary field $\phi_{n,m}$ has the conformal dimension:

$$\Delta_{n,m} = \frac{(np - mq)^2 - (p-q)^2}{4pq}, \quad 1 \le n \le q-1, \quad 1 \le m \le p-1. \qquad (A.5)$$

Here $\phi_{n,m}$ and $\phi_{q-n,p-m}$ have the same conformal dimension and are identified. The minimal model $M_{p,q}$ model contains $\frac{1}{2}(p-1)(q-1)$ primary fields.

In the minimal models, a set of models $M_{p,p+1}$ ($p = 3, 4, \ldots$) are called the **unitary series**, where all $\Delta_{n,m}$ are non-negative. For example, $M_{3,4}$ contains three primary fields $\phi_{1,1} = I$ (identity operator), $\phi_{1,2}$ with $\Delta_{1,2} = \frac{1}{16}$ and $\phi_{1,3}$ with $\Delta_{1,3} = \frac{1}{2}$. This model is called the Ising model whose central charge is $c = \frac{1}{2}$. Non-unitary models $M_{p,q}$ contain primary fields with negative conformal dimensions. For example, $M_{2,5}$ is the non-unitary minimal model which contains two primary fields $\phi_{1,1} = I$ and $\phi_{1,2}$ with $\Delta_{1,2} = -\frac{1}{5}$. This model is called the Yang–Lee edge singularity model whose central charge is $c = -\frac{22}{5}$.

It is useful to study the minimal model by using a free boson $\varphi(z)$, where the energy–momentum tensor is represented by

$$T(z) = -\frac{1}{2} : (\partial\varphi)^2 : + i\alpha_0 \partial^2 \varphi. \tag{A.6}$$

Here the symbol ": :" implies the normal ordering and $\alpha_0 = \sqrt{2}\left(\beta - \frac{1}{\beta}\right)$. Then the central charge is given by

$$c = 1 - 6\left(\beta - \frac{1}{\beta}\right)^2. \tag{A.7}$$

For $\beta = \sqrt{\frac{q}{p}}$, one obtains the central charge of the minimal model. The primary field of the CFT is realized by a vertex operator

$$V_\alpha(z) =: \exp(i\sqrt{2}\alpha\varphi(z)) : . \tag{A.8}$$

It has a conformal dimension

$$\Delta = \alpha^2 - 2\alpha_0\alpha. \tag{A.9}$$

For $\alpha = -\frac{1}{2}(1 - m)\frac{1}{\beta} + \frac{1}{2}(1 - m)\beta$, one obtains the conformal dimension (A.5) of the primary field $\phi_{n,m}$.

For a non-unitary minimal model, there is a primary field with negative conformal weight. We define the **effective central charge**

$$c_{\text{eff}} = c - 24\Delta_{\min}. \tag{A.10}$$

Here $\Delta_{\min}$ is the lowest conformal dimension among the primary fields in the CFT. For the minimal model $M_{p,q}$, the lowest conformal dimension is

$$\Delta_{\min} = \frac{1 - (p - q)^2}{4pq}. \tag{A.11}$$

Then the effective central charge is given by

$$c_{\text{eff}} = 1 - \frac{6}{pq}. \tag{A.12}$$

A.1.1 BLZ Approach

CFT is realized as the IR or UV fixed point of the renormalization group flow in a space of coupling constants of a field theory. Let us consider a field theory that is obtained by the continuum limit of an integrable lattice model. Such a theory is expected to have an integrable structure. A notable example is the UV CFT perturbed by a relevant perturbation, which flows to the IR field theory. Intermediate effective field theory is described by a set of pseudo-particles with S-matrices constrained by integrability. CFT is expected to possess such an integrable structure. Integrability guarantees the existence of infinite commuting conserved charges. The composite operators formed by normal ordered powers of stress tensor and its derivative give rise to an infinite-dimensional commuting subalgebra spanned by the local integrals of motion [4]. This integrability structure can be also described by the massless S-matrix. In a series of papers [5–8], Bazhanov, Lukyanov and Zamoldchikov (BLZ) studied the integrability structure and functional relations of the minimal CFT.

In the case of CFT with central charge $c < 1$, an infinite number of the local integral of motion leads to the KdV hierarchy after taking the classical limit $c \rightarrow -\infty$ and replacing the commutator by the Poisson brackets [5]. One conventional way to generate this classical KdV hierarchy is by using the Lax pair and the monodromy matrix, which are spanned by the generators of $sl(2)$ Lie algebra. In particular, the expansion of the transfer operator, the trace of the monodromy operator, leads to infinite classical integrals of motion.

By introducing the quantum universal enveloping algebra $U_q(sl(2))$ [5], the quantum counterparts of the monodromy matrix are defined, which leads to the operator-valued matrices called the monodromy operator. This operator is designed to satisfy the quantum Yang–Baxter equation with a trigonometric R matrix. The transfer operator is defined as the trace of the monodromy operators. The transfer matrix operators can be regarded as the generating function of the non-local integrals of motion, which commute among themselves and commute with the local integrals of motion. The eigenfunctions of the transfer operators are an entire function of the spectral parameter λ^2. Similarly to the classical transfer matrix, the asymptotic of the transfer operator at $\lambda^2 \rightarrow \infty$ is related to the local integrals of motion [5, 8]

$$\mathbf{T}(\lambda) = \Lambda(q\lambda) + \Lambda^{-1}(q^{-1}\lambda),$$

$$\log \Lambda(q\lambda) \simeq m\lambda^{1+\xi} - \sum_{k=1}^{\infty} C_k I_{2k-1} \lambda^{(1-2k)(1+\xi)}, \tag{A.13}$$

where m and C_n are defined by

$$m = \frac{2\sqrt{\pi}\Gamma(\frac{1}{2} - \frac{\xi}{2})}{\Gamma(1 - \frac{\xi}{2})}[\Gamma(\frac{1}{1+\xi})]^{1+\xi},$$

$$C_n = \frac{1+\xi}{n!}(\frac{\pi\xi}{1+\xi})^n[\frac{2\Gamma(\frac{1}{2} - \frac{\xi}{2})}{m\Gamma[1 - \frac{\xi}{2}]}]^{2n-1}\frac{\Gamma[(n - \frac{1}{2})(1+\xi)]}{\Gamma[1 + (n - \frac{1}{2})\xi]}.$$

(A.14)

ξ is defined by the central charge c and conformal dimension Δ

$$\beta = \sqrt{\frac{1-c}{24}} - \sqrt{\frac{25-c}{24}}, \quad \xi = \frac{\beta^2}{1 - \beta^2},$$

$$\Delta = (\frac{p}{\beta})^2 + \frac{c-1}{24}.$$

(A.15)

In the case of a minimal model $M_{2,2n+3}$, ξ is given by

$$c = 1 - 3\frac{(2n+1)^2}{2n+3}, \quad \xi = \frac{2}{2n+1}.$$

(A.16)

According to the ODE/IM correspondence, the Stokes multiplier, T-function, $T_1(E)$ corresponds to the eigenvalue of the vacuum state of $\mathbf{T}(\lambda)$, where the first few eigenvalues of I_{2k-1} is

$$I_1^{vac}(\Delta) = \Delta - \frac{c}{24},$$

$$I_3^{vac}(\Delta) = \Delta^2 - \frac{c+2}{12}\Delta + \frac{c(5c+22)}{2880},$$

$$I_5^{vac}(\Delta) = \Delta^3 - \frac{c+4}{8}\Delta^2 + \frac{(c+2)(3c+20)}{576}\Delta - \frac{c(3c+14)(7c+68)}{290304},$$

$$I_7^{vac}(\Delta) = \Delta^4 - \frac{c+6}{6}\Delta^3 + \frac{15c^2 + 194c + 568}{1440}\Delta^2$$
$$- \frac{(c+2)(c+10)(3c+28)}{10368}\Delta + \frac{c(3c+46)(25c^2 + 426c + 1400)}{24883200}.$$

(A.17)

In the case of the minimal model $M_{2,2n+3}$, the transfer operators satisfy the T-system, which is identical to the functional relation satisfied by the transfer matrices of the XXZ model [9–11]. From the T-system one furthermore finds the Y-system and TBA equations. Interestingly, the Y-system (the TBA equations resp.) have the same form as those derived from the Schrödinger equation in Sect. 1.1 in the main text. It would be a very critical test of the ODE/IM correspondence to compare the asymptotic behavior of the Y-function obtained from the BLZ approach with the one obtained from the Schrödinger equation.

We first fix the dictionary of the parameters on both sides by comparing the asymptotic behavior of Q-functions (1.83) in the main text from the Schrödinger equation at large E with the one at large λ derived in [6, 12][1]

$$\log Q \sim \frac{\Gamma(\frac{\xi}{2})\Gamma(\frac{1}{2} - \frac{\xi}{2})}{\sqrt{\pi}}[\Gamma(1 - \beta^2)]^{1+\xi}(-\lambda^2)^{\frac{1}{2-2\beta^2}}. \tag{A.18}$$

The identifications between the two sides are found to be

$$2\mu = 1 + \xi, \quad \beta^2 = 1 - \frac{1}{2\mu}, \quad \mu = \frac{M+1}{2M},$$

$$E = e^{\theta/\mu} = [2(M+1)]^{1/\mu}[\Gamma(\frac{1}{2\mu})]^2\lambda^2. \tag{A.19}$$

Let us focus on the simplest case, the Lee–Yang mode $M_{2,5}$, where we can find the expansion of the T-function by (A.13). Let us denote the expansion of $\log Y_2(E)$, where $Y_2(E) = T(E)$, by

$$\log Y_2(E) = me^\theta + \sum_{n=1} m^{(2n)}e^{(1-2n)\theta}. \tag{A.20}$$

According to the ODE/IM correspondence, this expansion should correspond to the expansion of the WKB period of the Schrödinger-type ODE

$$\left(-\frac{d^2}{dz^2} + z^3 - E\right)\psi = 0. \tag{A.21}$$

After rescaling $x = \hbar^{\frac{2}{5}}z$ and $v = \hbar^{\frac{6}{5}}E$, one finds

$$\left(-\hbar^2\partial_x^2 + x^3 - v\right)\psi(x) = 0. \tag{A.22}$$

The Y-function Y_2 is expected to coincide with the WKB period

$$\Pi_\gamma = 2\int_{x_1}^{x_2} P(x)dx = \sum_{n=0}^\infty \hbar^{2n}\Pi^{(n)}, \tag{A.23}$$

where $P(x)$ is defined as $P_{\text{even}}(x)$ in (2.5) in the main text. x_1 and x_2 are the left two turning points. One can check that the mass m in (A.20) coincides with the classical WKB period $\Pi_\gamma^{(0)}$. By using the differential operator method presented in Sect. 2.1, we can compute the higher-order corrections of the WKB period Π_γ:

---

[1] See also [13].

$$\Pi_\gamma = \sum_{n=0}^{\infty} \left(\frac{\nu}{E}\right)^{\frac{5n}{3}} \Pi^{(n)}. \tag{A.24}$$

Comparing the $\log Y_2$ expansion and the Π expansion numerically, one finds

$$m^{(2n)} = \nu^{\frac{5n}{3}} \Pi^{(n)} \quad n = 1, 2 \ldots . \tag{A.25}$$

Moreover, it has been shown that the integrals of motion I_{2k-1} of $M_{2,5}$ vanish whenever $2k - 1$ is divisible by 3 [14], which implies

$$m_4 = m_{10} = m_{16} = \ldots = 0. \tag{A.26}$$

This is also easily confirmed by using the differential operator method of Schrödinger equation. This thus provides a non-trivial test to the ODE/IM correspondence.

A.2   Integrable Field Theory and TBA Equations

The field theory away from the RG fixed point becomes a massive field theory with highly nontrivial interaction terms, which is difficult to study in general. However, an integrable field theory that has an infinite number of conserved charges is strongly constrained. One can compute the mass spectrum, the S-matrix of the scattering, and thermodynamics.

A.2.1   S-Matrix and Integrability

Let us consider the thermodynamics of (1+1)-dimensional massive integrable quantum field theories, which have an infinite number of commuting integrals of motions. Such integrable field theories in the infrared region can be described by factorizable S-matrices among the set of massive particles. We parametrize the energy and momentum of a particle with mass m by rapidity θ:

$$(p^0, p^1) = (m \cosh \theta, m \sinh \theta). \tag{A.27}$$

We assume that the S-matrix takes the form of diagonal scattering. Namely, in the 2-body scattering process, there is no reflection. This happens only for an elastic scattering process. In this case, the scattering matrix S_{ab}^{cd} from the states a and b to c and d takes the form

$$S_{ab}^{cd}(\theta) = \delta_a^c \delta_b^d S_{ab}(\theta). \tag{A.28}$$

$S_{ab}(\theta)$ represents the scattering amplitude of two particles a and b, where θ is the relative rapidity $\theta_{ab} = \theta_a - \theta_b$. An infinite number of conserved charges consistent with the S-matrix impose strong constraints on the amplitude. Assuming also the unitarity, reflection symmetry, and analyticity, the scattering amplitudes are shown to be factorized forms. The scattering amplitude S_{ab} has a simple pole at a certain value of θ_{ab}, which corresponds to the bound state of particles a and b. The factorization properties imply that the scattering amplitudes satisfy the **bootstrap equation**. This equation also restricts the structure of the S-matrix. Such a theory is called **purely elastic scattering theory**.

A simple example of this type of field theory is the **Yang–Lee edge singularity** [15], which is realized as the renormalization group fixed point of the Ising model with an imaginary-valued magnetic field. The theory contains a single massive particle with mass m. The S-matrix is given by

$$S(\theta) = \frac{\sinh\theta + i\sin\frac{\pi}{3}}{\sinh\theta - i\sin\frac{\pi}{3}}. \tag{A.29}$$

In the purely elastic scattering theory, one can deal with the wave functions with definite particle numbers, since no new particles are created or annihilated, and all the momenta are conserved in the scattering process. For the Yang–Lee theory, let us consider the N-particles located at positions $x_1, \ldots, x_N$ and denote its wave function by $\Psi(x_1, \ldots, x_N)$. The wave function is supposed to be (anti-)periodic:

$$\Psi(x_1, \ldots, x_i + L, \ldots, x_N) = \pm\Psi(x_1, \ldots, x_N). \tag{A.30}$$

When all the particles are well-separated compared to the characteristic scale ξ of the theory, the particles are free. For $x_{i_1} \ll x_{i_2} \ll \cdots \ll x_{i_N}$, the wave function becomes

$$\Psi(x_1, \ldots, x_N) = \prod_{i=1}^{N} e^{ip_{i_k}x_{i_k}}. \tag{A.31}$$

When the ordering of the positions x_{i_p} and $x_{i_{p+1}}$ is changed, the two wave functions are related by multiplying the S-matrix $S(\theta_{i_p} - \theta_{i_{p+1}})$. Now we consider the case that x_i moves around the circle and goes back to the original position. The total changes of the wave function should be equal to the boundary condition (A.30). We then obtain the **Bethe ansatz equation (BAE)**:

$$e^{ip_i L} \prod_{j \neq i} S(\theta_i - \theta_j) = \pm 1. \tag{A.32}$$

Here the plus (minus) sign refers to the periodic (anti-periodic) boundary condition. Taking the logarithm of (A.32), we obtain a different form of the BAE:

$$mL \sinh \theta_i + \sum_{j \neq i} \varphi(\theta_i - \theta_j) = 2n_i \pi, \tag{A.33}$$

where n_i is a (half-)integer for (anti-)periodic boundary conditions and $\varphi(\theta) = -i \log S(\theta)$. For a solution to the BAE, one obtains the total energy and momentum of the system:

$$E = \sum_{i=1}^{N} m \cosh \theta_i, \quad P = \sum_{i=1}^{N} m \sinh \theta_i. \tag{A.34}$$

A.2.2 TBA Equations

Now we want to consider the thermodynamics of such a theory. We consider a purely elastic scattering theory defined on a circle. We impose the periodic boundary condition in both space and time directions. The spacetime geometry becomes the torus. Since we are dealing with Euclidean field theories, we can think of time directions in many ways. Let us consider the partition function of the model on the torus obtained by identifying a rectangle with lengths L and R (Fig. A.1).

When we think of L-direction as time and R-direction as space, the partition function is defined as

$$Z(R, L) = \mathrm{Tr} e^{-LH_R}, \tag{A.35}$$

where H_R is the Hamiltonian defined on the circle. The trace is taken over the Hilbert space of the states. On the other hand, when we regard the R-direction as the time, the partition function is given by

$$Z(R, L) = \mathrm{Tr} e^{-RH_L}. \tag{A.36}$$

Here H_L denotes the Hamiltonian defined on the L-direction.

The partition function (A.35) describes the system in the inverse temperature L. The ground state energy $E_0(R)$ can be read off from the behavior in the low-temperature limit $L \to \infty$:

Fig. A.1 Torus obtained by identifying two sides of the rectangle

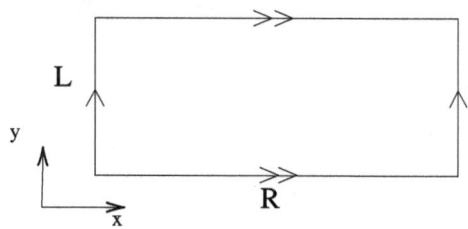

$$Z(R, L) \sim e^{-LE_0(R)}, \quad (L \to \infty). \tag{A.37}$$

The logarithm of the partition function (A.35) is the free energy $F(L, R)$ in the inverse temperature R:

$$Z(R, L) = e^{-RF(L,R)}. \tag{A.38}$$

Let $f(R)$ denote the free energy per unit length $F(L, R) = Lf(R)$. Then we find the relation

$$E_0(R) = Rf(R). \tag{A.39}$$

Since the free energy is determined by the TBA equation, we can estimate the behavior of the ground state energy. Let $T_{\mu\nu}$ be the energy–momentum tensor of the theory, where we take x as the R-direction and y as the L-direction, respectively. The Hamiltonian H_R in the "R-time" picture is expressed as

$$H_R = \frac{1}{2\pi} \int T_{yy} dx. \tag{A.40}$$

The Hamiltonian H_L is given by

$$H_L = \frac{1}{2\pi} \int T_{xx} dy. \tag{A.41}$$

The vacuum expectation value of the energy density T_{yy} is obtained from

$$\frac{1}{2\pi} \int dx \langle T_{yy} \rangle = \frac{\mathrm{Tr} H_R e^{-LH_R}}{\mathrm{Tr} e^{-LH_R}} = -\frac{d}{dL} \log \mathrm{Tr} e^{-LH_R} \sim E_0(R). \tag{A.42}$$

We find

$$\langle T_{yy} \rangle = 2\pi \frac{E_0(R)}{R}. \tag{A.43}$$

On the other hand, in the "L-time" picture, the vacuum expectation value (VEV) of the T_{xx} satisfies

$$\frac{L}{2\pi} \langle T_{xx} \rangle = -\frac{d}{dR} \mathrm{Tr} e^{-RH_L} = L\frac{d}{dR} E_0(R), \tag{A.44}$$

from which we obtain

$$\langle T_{xx} \rangle = 2\pi \frac{dE_0(R)}{dR}. \tag{A.45}$$

For the VEV of the trace of the energy–momentum tensor $\Theta = T_{xx} + T_{yy}$, we find

$$\langle\Theta\rangle = \frac{2\pi}{R}\frac{d}{dR}(RE_0(R)). \tag{A.46}$$

We parametrize $E_0(R)$ as

$$E_0(R) = -\frac{\pi\tilde{c}(r)}{E}, \tag{A.47}$$

where $r = m_1 R$ and m_1 is the lowest mass. r is a dimensionless parameter. The function $\tilde{c}(r)$ is called the **scaling function**. In the UV limit where $r \to 0$, the theory goes to a conformal field theory. Then the VEV of the trace of the energy–momentum tensor corresponds to the effective central charge of the CFT. In this limit, the scaling function $\tilde{c}(r)$ approaches the effective central charge c_{eff}. In particular, the Yang–Lee edge singularity $\mathcal{M}_{2,5}$ has the central charge $c = -\frac{22}{5}$ and the effective central charge $c_{\text{eff}} = \frac{2}{5}$.

We will compute the free energy of the system based on the microscopic theory with purely elastic scattering S-matrix. Here we focus on the Yang–Lee edge singularity again. The thermodynamic limit is defined as the large size $L \to \infty$ and the large number of particles $N \to \infty$ while the density N/L is kept finite. The solutions $\{\theta_i\}$ corresponding to the N-particles define a set of integers $\{n_i\}$, which are called the roots. The complement set of integers corresponds to the holes. In the thermodynamic limit, the distribution of the roots is characterized by the density $\rho^{(r)}(\theta)$, where $\rho^{(r)}(\theta)\Delta\theta$ represents the number of roots in the region $\theta \sim \theta + \Delta\theta$ per unit length of the system. The hole density $\rho^{(h)}(\theta)$ is defined similarly. Using the density of roots, the BAE (A.33) can be written as

$$m\sinh\theta_i + \int\varphi(\theta_i - \theta')\rho^{(r)}(\theta')d\theta' = \frac{2\pi n_i}{L}. \tag{A.48}$$

We introduce the **counting function**

$$J(\theta) = \frac{m}{2\pi}\sinh\theta + \frac{1}{2\pi}\varphi * \rho^{(r)}(\theta). \tag{A.49}$$

The function $LJ(\theta)$ takes values in integers for the roots and holes. The total density of the holes and roots is obtained by

$$\rho(\theta) := \rho^{(r)}(\theta) + \rho^{(h)}(\theta) = \frac{dJ(\theta)}{d\theta}. \tag{A.50}$$

Using the total and the root densities, the total energy and the entropy of the system are obtained as

$$E[\rho^{(r)}] = \int_{-\infty}^{\infty} m \cosh \theta \rho^{(r)}(\theta), \tag{A.51}$$

$$S[\rho, \rho^{(r)}] = \pm \int_{-\infty}^{\infty} d\theta \left[\rho \log \rho \mp \rho^{(r)} \log \rho^{(r)} - \log(\rho \pm \rho^{(r)}) \log(\rho \pm \rho^{(r)}) \right], \tag{A.52}$$

where the plus sign corresponds to the fermionic case and the minus to the bosonic case. The free energy of the system can be obtained by

$$f[\rho, \rho^{(r)}] = E[\rho^{(r)}] - T S[\rho, \rho^{(r)}]. \tag{A.53}$$

Minimizing the free energy for $\rho^{(r)}$, we find the free energy for the ground state. It is convenient to introduce the **pseudo energy** $\epsilon(\theta)$ by

$$\frac{\rho^{(r)}(\theta)}{\rho(\theta)} = \begin{cases} \frac{e^{-\epsilon}}{1+e^{-\epsilon}} & \text{fermionic} \\ \frac{e^{-\epsilon}}{1-e^{-\epsilon}} & \text{bosonic} \end{cases}. \tag{A.54}$$

The extremal condition is written as

$$m R \cosh \theta = \epsilon(\theta) \pm \frac{1}{2\pi} \varphi * \log(1 \pm e^{-\epsilon})(\theta). \tag{A.55}$$

This equation is called the **Thermodynamic Bethe Ansatz (TBA) equation**, which determines the pseudo energy $\epsilon(\theta)$ from the S-matrix. The free energy is obtained as

$$f(R) = \mp \frac{1}{R} \int_{-\infty}^{\infty} m \cosh \theta \log(1 \pm e^{-\epsilon(\theta)}) \frac{d\theta}{2\pi}. \tag{A.56}$$

For the Yang–Lee edge singularity, the S-matrix satisfies the fermionic condition $S(0) = -1$. Then we take the plus sign in (A.55). From (A.29), the kernel function $\varphi(\theta)$ is found to be

$$\varphi(\theta) = -\frac{4\sqrt{3}\cosh(\theta)}{1 + 2\cosh(2\theta)}. \tag{A.57}$$

The ground state energy is determined by (A.39), from which $\tilde{c}(r)$ is evaluated as

$$\tilde{c}(r) = \frac{3}{\pi^2} r \int_{-\infty}^{\infty} d\theta L(\theta) \cosh \theta, \tag{A.58}$$

where $L(\theta) = \log(1 + e^{-\epsilon(\theta)})$. The scaling function $\tilde{c}(r)$ can be calculated exactly in the IR ($r \to \infty$) and the UV ($r \to 0$) limits. The profile of the $L(\theta) = \log \left(1 + \right.$

Fig. A.2 Patterns of
$L(\theta) = \log\left(1 + e^{-\epsilon(\theta)}\right)$ for
different values of $r = mR$

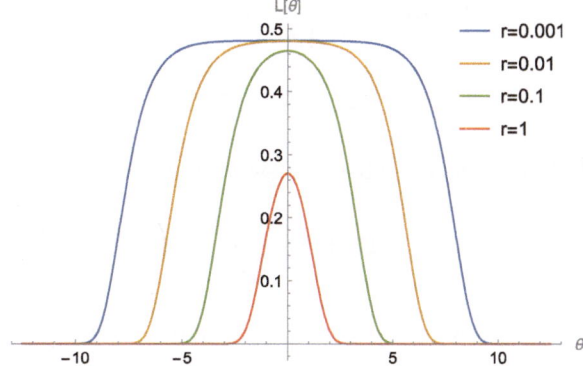

$e^{-\epsilon(\theta)}$) is seen in Fig. A.2. It has a plateau in the region $-\log\frac{2}{r} < \theta < \log\frac{2}{r}$. It becomes constant in the UV limit $r \to 0$. The limit $\epsilon = \lim_{r \to 0} \epsilon(\theta)$ satisfies the equation

$$\epsilon - N\log(1 + e^{-\epsilon}) = 0, \tag{A.59}$$

where

$$N = -\frac{1}{2\pi}\int_{-\infty}^{\infty} d\theta\varphi(\theta) = 1. \tag{A.60}$$

The Eq. (A.59) for ϵ is solved as $\epsilon_0 = \log(\frac{1+\sqrt{5}}{2})$.

Now we consider the TBA equation in the limit $r \to 0$. From the profile of the solution, the two edges (the right and left kinks) of the profile of $\epsilon(\theta)$ contribute to the integral. To evaluate the contribution from the right kink where θ is around $\log\frac{2}{r}$, we shift the parameter $\theta \to \theta - \log\frac{2}{r}$. The driving term $r\cosh\theta \to e^{\theta - \log\frac{2}{r}}$. Then the TBA equation reduces to

$$\epsilon(\theta) = e^{\theta} - \frac{1}{2\pi}\varphi * L(\theta). \tag{A.61}$$

This TBA equation in the kink limit plays an important role in the analysis of the ODE/IM correspondence. In the kink limit, the scaling function is evaluated as

$$\tilde{c}(0) = 2 \times \frac{3}{\pi^2}\int_{-\infty}^{\infty} L(\theta)e^{\theta}d\theta. \tag{A.62}$$

Here the factor 2 comes from contributions from two kinks. Substituting the θ-derivative of (A.61) into (A.62), we find

$$\tilde{c}(0) = \frac{6}{\pi^2} \left\{ \int_{\epsilon_0}^{\infty} \log(1 + e^{-x}) dx + \frac{1}{2}\epsilon_0 \log(1 + e^{-\epsilon_0}) \right\}$$

$$= \frac{6}{\pi^2} \mathcal{L}(\frac{1}{1 + e^{\epsilon_0}}). \tag{A.63}$$

Here $\mathcal{L}(x)$ is the Rogers dilogarithm function

$$\mathcal{L}(x) = -\frac{1}{2} \int_0^x dy \left(\frac{\log y}{1 - y} + \frac{\log(1 - y)}{y} \right). \tag{A.64}$$

For the value $\epsilon_0 = \log((1 + \sqrt{5})/2)$, the scaling function $\tilde{c}(0)$ is evaluated as

$$\tilde{c}(0) = \frac{2}{5}. \tag{A.65}$$

This agrees with the effective central charge of the Yang–Lee edge singularity.

A.2.3   Y-System

The previous integrable field theory with a single particle can be generalized to the theories with multi-particles. Here we discuss the integrable field theories with a single mass scale and multiple mass scales. The former example is the integrable theories whose diagonal scattering Š-matrix is characterized by simple Lie algebras [16]. The latter example we discuss here is the homogeneous sine-Gordon model. Both examples appear in the study of the ODE/IM correspondence. We will start the description of the TBA system for these theories as we did for the Yang–Lee edge singularity in the previous section. We can also begin with the functional relations in a more abstract way. The relations are nothing but the Y-system which was explained in Sect. 1.1.

Minimal Purely Elastic Scattering Theories

The diagonal scattering theories with minimal S-matrices are characterized by simply-laced Lie algebra. Let $\mathfrak{g}$ be a simply-laced Lie algebra of rank r. We denote h as the Coxeter number of $\mathfrak{g}$ and I_{ab} $(a, b = 1, \ldots, r)$ the incidence matrix defined by $2\delta_{ab} - A_{ab}$, where A_{ab} is the Cartan matrix of $\mathfrak{g}$. The theory contains r particles with mass m_a $(a = 1, \ldots, r)$ and the diagonal scattering matrices $S_{ab}(\theta)$ [16–18], where the S-matrices are represented in a factorized form. The masses of the particles obey the linear relation:

$$\sum_{b=1}^{r} I_{ab} m_b = 2 m_a \cos\left(\frac{\pi}{h}\right), \tag{A.66}$$

where $(m_1, \ldots, m_r)$ is an eigenvector of the incidence matrix I, whose entries are non-negative integers. This vector is called the **Perron–Frobenius eigenvector**. The TBA equations are expressed as [19]

$$\epsilon_a(\theta) = m_a R \cosh\theta - \sum_{b=1}^{r} \varphi_{ab} * L_b(\theta), \tag{A.67}$$

where

$$\varphi_{ab}(\theta) = -i \frac{d}{d\theta} \log S_{ab}(\theta), \tag{A.68}$$

and

$$L_b(\theta) = \log(1 + e^{-\epsilon_b(\theta)}). \tag{A.69}$$

A simple feature of the S-matrices is that $S_{ab}(\theta)$ is an analytic function of θ in the strip $0 < \mathrm{Im}\theta < \pi$ in the complex θ-plane and has simple poles in the strip. They factorize in the fundamental blocks which are responsible for the poles. Moreover, the Fourier transform of the $\varphi_{ab}(\theta)$ defined by

$$\tilde{\varphi}_{ab}(k) = \int_{-\infty}^{\infty} \varphi_{ab}(\theta) e^{ik\theta} d\theta \rightarrow \mathcal{F}[\varphi_{ab}](k) = \int_{-\infty}^{\infty} \varphi_{ab}(\theta) e^{-ik\theta} d\theta \tag{A.70}$$

satisfies the relation [19, 20]

$$\left(\delta_{ab} - \frac{1}{2\pi} \mathcal{F}[\varphi_{ab}](k)\right)^{-1} = \delta_{ab} - \frac{1}{2\cosh\frac{k}{h}} I_{ab}. \tag{A.71}$$

The relations (A.66) and (A.71) are used to rewrite the TBA Eq. (A.67) into the form

$$\epsilon_a(\theta) = m_a R \cosh\theta + \frac{1}{\pi} \sum_{b=1}^{r} I_{ab} \varphi_h * \left(\log(1 + e^{\epsilon_b}) - m_b R \cosh\right)(\theta), \tag{A.72}$$

where $\varphi_h(\theta)$ is defined by

$$\varphi_h(\theta) = \frac{1}{2\pi} \int_{-\infty}^{\infty} \frac{e^{-ik\theta}}{2\cosh\frac{k}{h}} dk = \frac{h}{2\cosh\frac{h}{2}\theta}. \tag{A.73}$$

$\varphi_h(\theta)$ has a pole at $\theta = \frac{\pi i}{h}$ and shows a discontinuity across the line $\theta = \frac{\pi i}{h}$. Shifting the TBA equation by $\theta \to \theta \pm \frac{\pi i}{h}$, one finds

$$\epsilon_a(\theta \pm \frac{\pi i}{h} \mp i0_+) = m_a R \cosh(\theta \pm \frac{\pi i}{h})$$
$$+ \frac{1}{\pi} \sum_{b=1}^{r} I_{ab} \int d\theta' \varphi_h(\theta - \theta' \pm \frac{\pi i}{h} \mp i0_+)\Big(\log(1 + e^{\epsilon_b}) - m_b R \cosh \Big)(\theta').$$

(A.74)

Taking summation of the equations with $\pm$ sign, one finds

$$\epsilon_a(\theta + \frac{\pi i}{h} - i0_+) + \epsilon_a(\theta - \frac{\pi i}{h} + i0_+) - 2m_a R \cos(\frac{\pi}{h}) \cosh \theta$$
$$= \sum_{b=1}^{r} I_{ab} \int d\theta' \sum_{b=1}^{r} \delta(\theta - \theta')\Big(\log(1 + e^{\epsilon_b}) - m_b R \cosh \Big)(\theta'),$$

(A.75)

where we have used (2.69) to compute the summation of $\varphi_h(\theta - \theta' + \frac{\pi i}{h} - i0_+) + \varphi_h(\theta - \theta' - \frac{\pi i}{h} + i0_+)\Big)$. Furthermore, using the relation of the masses (A.66), one finds

$$\epsilon_a(\theta + \frac{\pi i}{h}) + \epsilon_a(\theta - \frac{\pi i}{h}) = \sum_{b=1}^{r} I_{ab} \log \left(1 + e^{\epsilon_b(\theta)}\right).$$

(A.76)

This leads to the functional equations for $Y_a(\theta) = e^{\epsilon_a(\theta)}$:

$$Y_a(\theta + i\frac{\pi}{h}) Y_a(\theta - i\frac{\pi}{h}) = \prod_{b=1}^{r}(1 + Y_b(\theta))^{I_{ab}}.$$

(A.77)

This defines the Y-system of ADE type [19]. The effective central charge for $\mathfrak{g} = A_r$ is $2r/(r+3)$, which is that of the $\mathbb{Z}_{r+1}$ parafermions.

Homogeneous Sine-Gordon Model

The above TBA system has a single mass scale, where mass parameters obey linear relations (A.66). The **homogeneous sine-Gordon model (HSG)** is defined as the gauged WZNW (Wess–Zumino–Novikov–Witten) model based on the coset space $G_k/U(1)^r$ perturbed by the potential term with multiple masses [21, 22]. Here G denotes a simple Lie group with rank r and k is the level of the WZNW model. The action of the model is formally written as

$$S_{\mathrm{HSG}} = S_{\mathrm{CFT}} + \frac{m^2}{\pi \beta^2} \int d^2 x \langle \Lambda_+, g(x)^{-1} \Lambda_- g(x) \rangle.$$

(A.78)

Here S_{CFT} denotes the action for the $G_k/U(1)^r$ CFT, or the level k G-parafermion, which depends on the G-valued field $g(x)$. $\langle\ ,\ \rangle$ is the Killing form on the Lie algebra $\mathfrak{g}$ of the Lie group G. m is the overall mass scale and β represents a coupling constant. $\Lambda_\pm$ are some semi-simple elements in the Cartan subalgebra of $\mathfrak{g}$, which are parameterized as

$$\Lambda_\pm = i\lambda_\pm \cdot H, \quad \lambda_\pm = \sum_{j=1}^r m_i e^{\pm\sigma_j} \omega_j. \tag{A.79}$$

Here H denotes the Cartan generators and ω_j the fundamental weights. σ_i is called the resonance parameter.

The particles of the effective theory are labeled by two indices (a, i) ($a = 1, \ldots, k-1, i = 1, \ldots, r$) with mass

$$M_a^i = M m_i \mu_a, \tag{A.80}$$

where M represents the overall mass scale, m_i are relative mass parameters, μ_a satisfies the condition (A.66). The S-matrices $S_{ab}^{ij}(\theta)$ of two particles of (a, i) and (b, j) are defined as follows [22–24]:

- For $i = j$, the S-matrix is determined by that of minimal purely elastic scattering theory for A_{k-1} (A.71):

$$S_{ab}^{ii}(\theta) = \exp\left\{\int_{-\infty}^{\infty} \frac{dt}{t} 2\cosh\frac{\pi t}{k} \left(2\cosh\frac{\pi t}{k} - I^{A_{k-1}}\right)_{ab}^{-1} e^{-it\theta}\right\}. \tag{A.81}$$

Here $I^{A_{k-1}}$ is the incidence matrix of A_{k-1}.
- For $i \neq j$, the scattering matrices are characterized by the Dynkin diagram associated with $\mathfrak{g}$ and the resonance parameters σ_i:

$$S_{ab}^{ij}(\theta) = \left[(\eta_{ij})^{ab} \exp\left(-\int \frac{dt}{t} \left(2\cosh\frac{\pi t}{k} - I^{A_{k-1}}\right)_{ab}^{-1} e^{-it(\theta+\sigma_{ij})}\right)\right]^{-I_{ij}^{\mathfrak{g}}}, \tag{A.82}$$

where $\sigma_{ij} = \sigma_i - \sigma_j$ and $I^{\mathfrak{g}}$ is the incidence matrix of $\mathfrak{g}$. $\eta_{ij} = \eta_{ji}^{-1}$ are some constant factors.

Due to the parity asymmetry, we need to treat the left-moving mode and the right-moving particles independently.

The TBA system is given by [23, 25]

$$\hat{\epsilon}^i_a(\theta) = M^i_a R \cosh\theta - \sum_{b=1}^{k-1}\left(\varphi_{ab} * \hat{L}^i_b - \sum_{j=1}^{r} I^{\mathfrak{g}}_{ij}\psi_{ab} * \hat{L}^j_b(\theta + \sigma_{ij})\right), \qquad (A.83)$$

where $I^{\mathfrak{g}}$ is the incidence matrix of $\mathfrak{g}$ and $\hat{L}^i_a = \log(1 + e^{-\hat{\epsilon}^i_a})$. $\varphi_{ab}(\theta)$ is the kernel for the minimal pure elastic scattering (A.68). The kernel function $\psi_{ab}(\theta)$ is defined in terms of the Fourier transformation:

$$\tilde{\psi}_{ab}(k) = \tilde{\varphi}_h(k)(\delta_{ab} - \frac{1}{2\pi}\tilde{\varphi}_{ab}(k)). \qquad (A.84)$$

The scaling function

$$c(R) = \frac{3}{\pi^2}\sum_{i=1}^{r}\sum_{a=1}^{k-1}\int d\theta M^i_a R\cosh\theta \hat{L}^i_a(\theta) \qquad (A.85)$$

is evaluated in the UV limit $R \to 0$, whose value is found to be

$$c_{\text{eff}} = \frac{k\dim\mathfrak{g}}{k+h} - r = \frac{k-1}{k+h}hr, \qquad (A.86)$$

where h is the Coxeter number of $\mathfrak{g}$. Introducing $\epsilon^i_a(\theta) = \hat{\epsilon}^i_a(\theta - \sigma_i)$, we can write the TBA system in the form

$$\epsilon^a_i(\theta) = v^a_i(\theta) - \sum_{b=1}^{k-1}\left(\varphi_{ab} * L^b_i - \sum_{j=1}^{r} I^{\mathfrak{g}}_{ij}\psi_{ab} * L^b_j\right)(\theta), \quad i = 1, \ldots, k-1, a = 1, \ldots, r,$$
$$(A.87)$$

where $L^a_i = \log(1 + e^{-\epsilon^a_i})$ and

$$v^a_i = \frac{1}{2}m^+_i \mu^a Re^{-\theta} + \frac{1}{2}m^-_i \mu^a Re^{\theta}, \qquad (A.88)$$

with $m^\pm_i = m_i e^{\pm\sigma_i}$. This is the $(A_{k-1}, \mathfrak{g})$-type TBA system[20]. Now the TBA equations take the universal form

$$\epsilon^a_i(\theta) = v^a_i(\theta) - \varphi_k * \left\{\sum_{b=1}^{k-1} I^{A_{k-1}}_{ab}(v^b_i - \log(1 + e^{\epsilon^b_i})) - \sum_{j=1}^{r} I^{\mathfrak{g}}_{ij}L^a_j\right\}(\theta), \quad (A.89)$$

where φ_k is given by (A.73). Then we can derive the Y-system for the HSG model $G_k/U(1)^r$, which takes the form of the $(A_{k-1}, \mathfrak{g})$-type Y-system

$$Y_a^i(\theta + \frac{i\pi}{k})Y_a^i(\theta - \frac{i\pi}{k}) = \prod_{b=1}^{k-1}(1 + Y_b^i(\theta))^{I_{ab}^{A_{k-1}}} \prod_{j=1}^{r}\left(1 + Y_a^j(\theta)^{-1}\right)^{-I_{ij}^{\mathfrak{g}}}, \quad (A.90)$$

where $Y_a^i(\theta) = e^{\epsilon_a^i(\theta)}$. In the case of (A_r, A_{hM-1}), we find the Y-system (3.73) of this type. The case where σ_i is pure imaginary numbers appeared in Chap. 2 and also the minimal surface in the AdS spacetime [26, 27].

A.3 Non-linear Integral Equation from Bethe Ansatz Equation

In this section, we derive the on-linear integral equation from the Bethe ansatz equation (1.35). We introduce the **counting function**

$$a(E) = \omega^{2\ell+1} \frac{Q(\omega^2 E)}{Q(\omega^{-2}E)} = \omega^{2\ell+1} \prod_{n=1}^{\infty} \frac{E_l - \omega^2 E_n}{E_l - \omega^{-2}E_n}. \quad (A.91)$$

The BAEs can be rewritten as

$$a(E_l) = -1, \quad l = 1, \ldots. \quad (A.92)$$

Taking the logarithm of the counting function, one obtains

$$\log a(E) = \frac{i\pi(2\ell + 1)}{M + 1} + \sum_{k=1}^{\infty} F(E/E_k), \quad (A.93)$$

where $F(E)$ is defined by

$$F(E) = \log\left(\frac{1 - \omega^2 E}{1 - \omega^{-2}E}\right). \quad (A.94)$$

Note that $1 + a(E)$ has a simple zero at E_l, where the logarithmic derivative $\partial_E \log\left(1 + a(E)\right)$ has a simple pole. We thus can rewrite the infinite sum as a contour integral

$$\log a(E) = \frac{i\pi(2\ell + 1)}{M + 1} + \int_C \frac{dE'}{2\pi i} F(E/E')\partial_{E'} \log\left(1 + a(E')\right), \quad (A.95)$$

where contour C encircling all the Bethe roots runs on a real axis. Changing variable via $E = e^{2M\theta/(M+1)^2}$ and integrating by parts, we obtain

---

[2] With a mild abuse of notation, $\log a(\theta) := \log a(e^{2M\theta/(M+1)})$.

$$\log a(\theta) = \frac{i\pi(2\ell+1)}{M+1} - \int_{C_1} d\theta' R(\theta-\theta') \log\left(1+a(\theta')\right)$$
$$+ \int_{C_2} d\theta' R(\theta-\theta') \log\left(1+a(\theta')\right),$$

(A.96)

where the function $R(\theta)$ is

$$R(\theta-\theta') = -\frac{1}{2\pi i}\partial_\theta F(\theta-\theta').$$

(A.97)

The circle C_1 (C_2) runs from $-\infty$ to ∞ just below (above) the real axis.

From the Bethe ansatz equation (1.35), it is easy to find

$$\left(a(\theta)\right)^* = a(\theta^*)^{-1},$$

(A.98)

which enables us to rewrite the integral in terms of the one along the real axis

$$\log a(\theta) = \frac{i\pi(2\ell+1)}{M+1} - 2i\int_{-\infty}^{\infty} d\theta' R(\theta-\theta')\text{Im}\log\left(1+a(\theta'-i0)\right)$$
$$+ \int_{-\infty}^{\infty} d\theta' R(\theta-\theta'-i0) \log\left(a(\theta'+i0)\right).$$

(A.99)

We then take the Fourier transform

$$\mathcal{F}[f(\theta)] = \int_{-\infty}^{\infty} d\theta f(\theta)e^{-ik\theta}$$

(A.100)

of both sides of (A.99) and then solve the equation about $\mathcal{F}\big[\log a\big](k)$. After that we perform the inverse Fourier transform to obtain the non-linear integral equation on $\log a(\theta)$

$$\log a(\theta) = -imLe^\theta + i\pi(\ell+\frac{1}{2}) + 2i\int_{-\infty}^{\infty} d\theta' \varphi(\theta-\theta')\text{Im}\big[1+a(\theta'-i0)\big],$$

(A.101)

which is simply called the **NLIE** [28, 29]. Here the kernel $\varphi(\theta)$ is given by

$$\varphi(\theta) = \mathcal{F}^{-1}\left[\frac{\mathcal{F}[R](k)}{1-\mathcal{F}[R](k)}\right](\theta) = \int_{-\infty}^{\infty} \frac{dk}{2\pi} e^{ik\theta} \frac{\sinh\left(\frac{\pi k(1-M)}{2M}\right)}{2\cosh\left(\frac{\pi k}{2}\right)\sinh\left(\frac{\pi k}{2M}\right)}.$$

(A.102)

The driving term $-imLe^\theta$ arises from the zero modes at $1 - \mathcal{F}[R](k) = 0$. To fix this term, we should compute the large E asymptotics of $Q(E)$ eq.(1.83), which leads to $mL = b_0/2$.

A.4 BAEs and NLIEs Associated with Simply-Laced Lie Algebras

So far we have studied the finite size of the integrable quantum field theory, which can be regarded as the continuous limit of the lattice models. The other way to investigate the finite size effects of the integrable field theories is the approach of the non-linear integral equations (NLIEs) [29–32], which starts from the (nested) BAEs of the XXZ spin chain and then takes the scaling limit.

Let us consider the rank r simply-laced Lie algebras $\mathfrak{g}$. The nested Bethe ansatz equations of the $\mathfrak{g}$-type generalization for the XXZ spin chain model read

$$
\left[\frac{\sinh\left(\gamma\left(\frac{h}{2\pi}(\lambda_{s,k} - \theta) + \frac{\pi i}{2}\right)\right) \sinh\left(\gamma\left(\frac{h}{2\pi}(\lambda_{s,k} + \theta) + \frac{\pi i}{2}\right)\right)}{\sinh\left(\gamma\left(\frac{h}{2\pi}(\lambda_{s,k} - \theta) - \frac{\pi i}{2}\right)\right) \sinh\left(\gamma\left(\frac{h}{2\pi}(\lambda_{s,k} + \theta) - \frac{\pi i}{2}\right)\right)} \right]^{M_{\delta_{s,1}}}
$$
$$
= \prod_{t=1}^{r} \prod_{j=1, j \neq k}^{M_s} \frac{\sinh\left(\gamma\left(\frac{h}{2\pi}(\lambda_{s,k} - \lambda_{s,j}) + \frac{i}{2}A_{st}\right)\right)}{\sinh\left(\gamma\left(\frac{h}{2\pi}(\lambda_{s,k} - \lambda_{s,j}) - \frac{i}{2}A_{st}\right)\right)},
$$
(A.103)

where A_{st} is the Cartan matrix of the Lie algebra $\mathfrak{g}$ and h the Coxter number. $\lambda_{s,k}$ is the rapidity of the s-th level vacuum with $k = 1, \ldots, M_s$. γ the anisotropy parameter and $\gamma h\theta/2\pi$ the inhomogeneity. The eigenvalues $e^{-iE_\pm}$ of the transfer matrix for $\lambda = \pm\theta$ are given by

$$
e^{-iE_\pm} = \prod_{k=1}^{M_1} \frac{\sinh\left(\gamma\left(\frac{h}{2\pi}(\mp\lambda_{1,k} + \theta) + \frac{i}{2}\right)\right)}{\sinh\left(\gamma\left(\frac{h}{2\pi}(\pm\lambda_{1,k} - \theta) + \frac{i}{2}\right)\right)}.
$$
(A.104)

The algebraic Bethe ansatz of the A-type was first studied in [33], and the D-type in [34]. When we take $\mathfrak{g} = A_1$, the BAEs become those of the standard XXZ spin chain.

The basic functions in the NLIEs are the counting function Z_s, $s = 1, \ldots, r$

$$
Z_s(\lambda) = M\delta_{s1}\left(\phi_{\frac{1}{2}}(\lambda - \theta) + \phi_{\frac{1}{2}}(\lambda + \theta)\right) - \sum_{k=1}^{M_s} \phi_1(\lambda - \lambda_{s,k}) + \sum_{t, A_{st}=-1} \sum_{k=1}^{M_t} \phi_{\frac{1}{2}}(\lambda - \lambda_{t,k}),
$$
(A.105)

where $\phi_\alpha(\lambda)$ is the odd function defined by

$$
\phi_\alpha = i \log \frac{\sinh\left(\gamma\left(\frac{h}{2\pi}\lambda + i\alpha\right)\right)}{\sinh\left(\gamma\left(-\frac{h}{2\pi}\lambda + i\alpha\right)\right)}.
$$
(A.106)

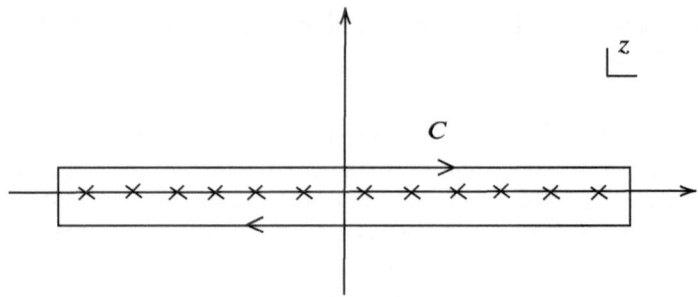

Fig. A.3 Contour C on complex z-plane. Here the crosses denote all the real roots and holes

In terms of the counting function, the BAEs can be written as

$$Z_s(\lambda_{s,k}) = 2\pi I_{s,k}, \qquad (A.107)$$

where $I_{s,k}$ is a half-integer. For the ground states, the Bethe roots are known to consist of the real roots. For an arbitrary low-lying excited state, one expects there are a large number of real roots denoted by $\rho_{s,k}, k = 1, \ldots, M_{R,s}$. For the excited states, there are also hole solutions satisfying $Z_s(\eta_{s,k}) = 2\pi I_{H,s,k}, k = 1, \ldots, M_{H,s}$. Moreover, all non-real Bethe roots called complex roots will be denoted by $\xi_{s,k}, k = 1, \ldots, M_{C,K}$. We also denote the other type of solution called "special" roots/holes satisfying $Z_s(\eta_{s,k}) = 2\pi I_{S,s,k}$ by $\sigma_{s,k}, k = 1, \ldots, M_{S,s}$ [31].

For arbitrary function f, one can use the trick

$$\oint_C \frac{dz}{2\pi i} f(z) \frac{d}{dz} \log\left(1 + (-1)^{\delta_s} e^{iZ_s(z)}\right) = \sum_{k=1}^{M_{R,s}} f(\rho_s, k) + \sum_{k=1}^{M_{H,s}} f(\eta_s, k), \quad (A.108)$$

where $\delta_s = r_s + M_s \bmod 2$ and $r_s = \delta_{1s} 2M - \sum_{t=1}^{r} C_{st} M_t$, to count all the real zeros of $1 + (-1)^{\delta_s} e^{iZ_s(z)}$. The contour C encircles all the real roots $\rho_{s,k}$, and the holes $\eta_{s,k}$. See Fig. A.3 for the contour C on the complex z-plane.

Applying this trick to the counting function and introducing the real function

$$Q_s(x) = -i \log \frac{1 + (-1)^{\delta_s} e^{iZ_s(x+i0)}}{1 + (-1)^{\delta_s} e^{-iZ_s(x-i0)}}, \qquad (A.109)$$

which satisfies $Q_s(x) = (Z_s(x) + \delta_s \pi) \bmod 2\pi$, one finds

$$\sum_{t=1}^{r} \mathcal{A}_{st} * (1 + \Phi_1) * \frac{d}{d\lambda} Z_t(\lambda) = 4\pi M \delta_{s1} \left(\Phi_{\frac{1}{2}}(\lambda - \theta) + \Phi_{\frac{1}{2}}(\lambda + \theta) \right)$$

$$+ \sum_{t=1}^{r} \left\{ \left(\mathcal{A}_{st} * (1 + \Phi_1) - 2\delta_{st} \right) * \frac{d}{d\lambda} Q_t \right.$$

$$- 2\pi \sum_{k=1}^{M_{H,t}} \left(\mathcal{A}_{st} * (1 + \Phi_1) - 2\delta_{st} \right)(\lambda - \eta_{t,k})$$

$$+ 2\pi \sum_{k=1}^{M_{C,t}} \left(\mathcal{A}_{st} * (1 + \Phi_1) - 2\delta_{st} \right)(\lambda - \xi_{t,k}) \right\},$$

$$(A.110)$$

where we have deformed the contour C such that it can be rewritten as integrals on the real axis. Here $\Phi_\alpha = \frac{1}{2\pi} \frac{d\phi_\alpha}{d\lambda}$ and $\mathcal{A}_{st}(\lambda)$ the generalized Cartan matrix defined by

$$\mathcal{A}_{st}(\lambda) = \begin{cases} 2\delta(\lambda) & s = t \\ -\frac{h}{2\pi} \frac{1}{\cosh(h\lambda/2)} & \Lambda_{st} = -1 \end{cases}.$$

$$(A.111)$$

We then multiply the inverse matrix of $\mathcal{A} * (1 + \Phi_1)$ and take the scaling limit $M \to \infty$, $\theta \to \infty$ while keeping $mL = Me^{-\theta}$ fixed. After integrating $\frac{d}{d\lambda} Z(\lambda)$ while taking care of the integration constant, the final equation reads [32]

$$Z_s = m_s L \sinh \lambda + \sum_{t=1}^{r} \left\{ X_{st} * Q_t + \sum_{k=1}^{M_{H,t}} \chi_{st}(\lambda - \eta_{t,k}) \right.$$

$$- 2 \sum_{k=1}^{M_{S,t}} \chi_{st}(\lambda - \sigma_{t,k}) - \sum_{k=1}^{M_{C,t}} \chi_{st}(\lambda - \xi_{t,k}) \right\},$$

$$(A.112)$$

where m_s are the masses of the solitons in the theory and form the Perron-Frobenius eigenvector (A.66). Here X_{st} are regular functions defined by

$$X_{st} = \delta_{st} - \frac{\sinh(\frac{\pi \kappa}{\gamma})}{\sinh(\kappa(\frac{\pi}{\gamma} - 1)) \cosh \kappa} \mathcal{A}_{st}^{-1}(\kappa)$$

$$(A.113)$$

and χ_{st} the odd primitive of $2\pi X_{st}$. $f(\kappa)$ is the Fourier transform of $f(\lambda)$

$$f(\kappa) = \int d\lambda e^{i\kappa h\lambda/\pi} f(\lambda).$$

$$(A.114)$$

When $\mathfrak{g}$ is A_r-type, the function $\mathcal{A}_{st}^{-1}(\kappa)$ reads

$$\mathcal{A}_{st}^{-1}(\kappa) = \coth\kappa \, \frac{\sinh(t\kappa)\sinh\big(\kappa(r+1-s)\big)}{\sinh\big(\kappa(r+1)\big)}, \quad s \geq t, \qquad (A.115)$$

and m_s is given by $\sin(\frac{\pi s}{r+1})$.

References

1. P. Di Francesco, P. Mathieu, D. Sénéchal, *Conformal Field Theory*. (Springer, 1997)
2. G. Mussardo, *Statistical Field Theory*. (Oxford Graduate Texts. Oxford University Press, 3, 2020)
3. A.A. Belavin, A.M. Polyakov, A.B. Zamolodchikov, Infinite conformal symmetry in two-dimensional quantum field theory. Nucl. Phys. **B241**, 333–380 (1984). https://doi.org/10.1016/0550-3213(84)90052-X
4. R. Sasaki, I. Yamanaka, Virasoro algebra, vertex operators, quantum Sine-Gordon and solvable quantum field theories. Adv. Stud. Pure Math. **16**, 271–296 (1988)
5. V.V. Bazhanov, S.L. Lukyanov, A.B. Zamolodchikov, Integrable structure of conformal field theory, quantum KdV theory and thermodynamic Bethe ansatz. Commun. Math. Phys. **177**, 381–398 (1996). https://doi.org/10.1007/BF02101898, arXiv:hep-th/9412229
6. V.V. Bazhanov, S.L. Lukyanov, A.B. Zamolodchikov, Integrable structure of conformal field theory. 2. Q operator and DDV equation. Commun. Math. Phys. **190**, 247–278 (1997). https://doi.org/10.1007/s002200050240, arXiv:hep-th/9604044
7. V.V. Bazhanov, S.L. Lukyanov, A.B. Zamolodchikov, Integrable quantum field theories in finite volume: Excited state energies. Nucl. Phys. B **489**, 487–531 (1997). https://doi.org/10.1016/S0550-3213(97)00022-9, arXiv:hep-th/9607099
8. V.V. Bazhanov, S.L. Lukyanov, A.B. Zamolodchikov, Integrable structure of conformal field theory. 3. The Yang-Baxter relation. Commun. Math. Phys. **200**, 297–324 (1999). https://doi.org/10.1007/s002200050531, arXiv:hep-th/9805008
9. A.N. Kirillov, N.Y. Reshetikhin, Exact solution of the integrable XXZ Heisenberg model with arbitrary spin. I. The ground state and the excitation spectrum. J. Phys. A **20**, 1565–1585 (1987). https://doi.org/10.1088/0305-4470/20/6/038
10. V.V. Bazhanov, N.Y. Reshetikhin, Critical Rsos models and conformal field theory. Int. J. Mod. Phys. A **4**, 115–142 (1989). https://doi.org/10.1142/S0217751X89000042
11. A. Klumper, P.A. Pearce, Conformal weights of RSOS lattice models and their fusion hierarchies. Physica A **183**, 304 (1992)
12. V.V. Bazhanov, S.L. Lukyanov, A.B. Zamolodchikov, Spectral determinants for Schrodinger equation and Q operators of conformal field theory. J. Statist. Phys. **102**, 567–576 (2001). https://doi.org/10.1023/A:1004838616921, arXiv:hep-th/9812247
13. P. Dorey, C. Dunning, R. Tateo, The ODE/IM Correspondence. J. Phys. A **40**, R205 (2007). https://doi.org/10.1088/1751-8113/40/32/R01, arXiv:hep-th/0703066
14. A. Maloney, G.S. Ng, S.F. Ross, I. Tsiares, Thermal Correlation Functions of KdV Charges in 2D CFT. JHEP **02**, 044 (2019). https://doi.org/10.1007/JHEP02(2019)044, arXiv:1810.11053 [hep-th]
15. A.B. Zamolodchikov, Thermodynamic Bethe Ansatz in relativistic models. Scaling Three State Potts and Lee-Yang Models. Nucl. Phys. **B342**, 695–720 (1990), https://doi.org/10.1016/0550-3213(90)90333-9
16. H.W. Braden, E. Corrigan, P.E. Dorey, R. Sasaki, Affine Toda field theory and exact S matrices. Nucl. Phys. **B338**, 689–746 (1990). https://doi.org/10.1016/0550-3213(90)90648-W

17. T.R. Klassen, E. Melzer, Purely elastic scattering theories and their ultraviolet limits. Nucl. Phys. B **338**, 485–528 (1990). https://doi.org/10.1016/0550-3213(90)90643-R
18. T.R. Klassen, E. Melzer, The Thermodynamics of purely elastic scattering theories and conformal perturbation theory. Nucl. Phys. B **350**, 635–689 (1991). https://doi.org/10.1016/0550-3213(91)90159-U
19. A.B. Zamolodchikov, On the thermodynamic Bethe ansatz equations for reflectionless ADE scattering theories. Phys. Lett. **B253**, 391–394 (1991). https://doi.org/10.1016/0370-2693(91)91737-G
20. F. Ravanini, R. Tateo, A. Valleriani, Dynkin TBAs.Int. J. Mod. Phys. **A8**, 1707–1728 (1993). https://doi.org/10.1142/S0217751X93000709, arXiv:hep-th/9207040 [hep-th]
21. C.R. Fernandez-Pousa, M.V. Gallas, T.J. Hollowood, J.L. Miramontes, The Symmetric space and homogeneous sine-Gordon theories. Nucl. Phys. B **484**, 609–630 (1997). https://doi.org/10.1016/S0550-3213(96)00603-7, arXiv:hep-th/9606032
22. J.L. Miramontes, C.R. Fernandez-Pousa, Integrable quantum field theories with unstable particles. Phys. Lett. B **472** 392–401 (2000). https://doi.org/10.1016/S0370-2693(99)01444-6, arXiv:hep-th/9910218
23. O.A. Castro-Alvaredo, A. Fring, C. Korff, J.L. Miramontes, Thermodynamic Bethe ansatz of the homogeneous Sine-Gordon models. Nucl. Phys. B **575**, 535–560 (2000). https://doi.org/10.1016/S0550-3213(00)00162-0, arXiv:hep-th/9912196
24. P. Dorey J.L. Miramontes, Mass scales and crossover phenomena in the homogeneous sine-Gordon models. Nucl. Phys. B **697**, 405–461 (2004). https://doi.org/10.1016/j.nuclphysb.2004.07.019, arXiv:hep-th/0405275
25. P. Dorey, J. Luis Miramontes, A T-duality interpretation of the relationship between massive and massless magnonic TBA systems. J. Stat. Mech. **0612**, P12016 (2006). https://doi.org/10.1088/1742-5468/2006/12/P12016, arXiv:hep-th/0609224
26. L.F. Alday, J. Maldacena, A. Sever, P. Vieira, Y-system for scattering amplitudes. J. Phys. A **43**, 485401 (2010). https://doi.org/10.1088/1751-8113/43/48/485401, arXiv:1002.2459 [hep-th]
27. Y. Hatsuda, K. Ito, K. Sakai, Y. Satoh, Thermodynamic Bethe Ansatz equations for minimal surfaces in AdS_3. JHEP **04**, 108 (2010). https://doi.org/10.1007/JHEP04(2010)108, arXiv:1002.2941 [hep-th]
28. A. Klümper, P.A. Pearce, Analytic calculation of scaling dimensions: Tricritical hard squares and critical hard hexagons. J. Stat. Phys. **64**(1), 13–76 (1991). https://doi.org/10.1007/BF01057867
29. C. Destri, H.J. de Vega, New thermodynamic Bethe ansatz equations without strings. Phys. Rev. Lett. **69**, 2313–2317 (1992). https://doi.org/10.1103/PhysRevLett.69.2313
30. C. Destri, H.J. De Vega, Unified approach to thermodynamic Bethe Ansatz and finite size corrections for lattice models and field theories. Nucl. Phys. **B438**, 413–454 (1995). https://doi.org/10.1016/0550-3213(94)00547-R, arXiv:hep-th/9407117 [hep-th]
31. C. Destri, H.J. de Vega, Nonlinear integral equation and excited states scaling functions in the sine-Gordon model. Nucl. Phys. B **504**, 621–664 (1997). https://doi.org/10.1016/S0550-3213(97)00468-9, arXiv:hep-th/9701107
32. P. Zinn-Justin, Nonlinear integral equations for complex affine Toda models associated to simply laced Lie algebras. J. Phys. A **31**, 6747–6770 (1998). https://doi.org/10.1088/0305-4470/31/31/019, arXiv:hep-th/9712222
33. O. Babelon, H.J. de Vega, C.M. Viallet, Exact Solution of the $Z(n+1) \times Z(n+1)$ symmetric generalization of the XXZ model. Nucl. Phys. B **200**, 266–280 (1982). https://doi.org/10.1016/0550-3213(82)90087-6
34. H.J. de Vega, M. Karowski, Exact Bethe Ansatz solution of 0(2n) symmetric theories. Nucl. Phys. **B280**, 225–254 (1987). https://doi.org/10.1016/0550-3213(87)90146-5

Index

© The Author(s), under exclusive licence to Springer Nature Singapore Pte Ltd. 2025
K. Ito and H. Shu, *ODE/IM Correspondence and Quantum Periods*, SpringerBriefs
in Mathematical Physics 51, https://doi.org/10.1007/978-981-96-0499-9